Darwin

**COMPETITION
&
COOPERATION**

Ashley Montagu

Darwin
COMPETITION
&
COOPERATION

GREENWOOD PRESS, PUBLISHERS
WESTPORT, CONNECTICUT

The Library of Congress has catalogued this publication as follows:

Library of Congress Cataloging in Publication Data

Montagu, Ashley, 1905-
    Darwin: competition & cooperation.

    Bibliography: p.
    1. Darwin, Charles Robert, 1809-1882.
2. Natural selection. I. Title.
[QH365.D8M6 1973]   595.01'62   72-11332
ISBN 0-8371-6657-8

To The Memory Of
PETR KROPOTKIN
1842-1921

Author of MUTUAL AID

CONTENTS

PREFACE

Untold numbers of people believe that what they understand by "Darwinism" has been scientifically demonstrated and is fully supported by the majority of contemporary scientists. "Darwinism," to most persons in the Western world today, usually connotes "the struggle for existence," "the survival of the fittest," "nature, red in tooth and claw," "the strongest survive, the weakest go to the wall," "eat or be eaten," "dog eat dog," "competition," and similar notions, all of them endowed with a content of combativeness—and all of them unsound.

The layman is not to be blamed for holding ideas that were largely conveyed to him, either directly or indirectly, by scientists who were themselves addicted to these erroneous ideas. The biologists of the nineteenth century, and many of their leading interpreters, were largely responsible for the spread of these erroneous notions, while, paradoxically enough, they were at the same time making the basic facts available upon which a sound theory of evolution could be developed. Unfortunately, what was sound and what was unsound in the theory of evolution became inextricably mixed for most people, and they either accepted the mixture or wholly rejected it. Meanwhile, biologists have been painstakingly illuminating the darker recesses of nature by their explorations of discovery, until today we have a theory of evolution based on such a wide (though by no means complete) assortment of related facts that most scientists can accept it without any reservations. The writer is among those who, with the majority of other scientists, accept the contemporary theory of evolution. It is very necessary to make this quite clear, for in this essay, in which we shall be concerned to discover how most people came to think the way they do about evolution, it will be necessary for the author to point out the weaknesses of certain aspects of Darwin's thought and work, weaknesses and errors that have had the most powerful and most unfortunate influences upon a very great number of human beings and their world.

This essay should be subtitled *The Darwinian Fallacy* for two reasons, firstly, because it attempts to exhibit a fundamental fallacy of Darwin's thought and that of his interpreters, and secondly, because it refers to the false ideas held by most persons concerning the meaning of Darwinism. Started on its course as something of a paradox, the Darwinian theory of evolution has for many years become almost a platitude.

To induce a better understanding of the significance of certain aspects of Darwinism, an endeavor is made here to indicate some of the sources of Darwin's fundamental ideas and modes of thought, and particularly to place Darwin's ideas in the social context in which they developed. The doctrine of "the survival of the fittest," whether correctly or incorrectly interpreted and applied, has done an untold amount of personal and social damage. This doctrine, which is essentially what has popularly come to be understood by Darwinism, is unsound. This essay attempts to indicate some of the sources as well as some of the grounds of this unsoundness.

After setting before the reader something of the history of the concept of natural selection as well as some of the fundamental misconceptions associated with it, the book will also provide an account of the theory of natural selection in its modern acceptable form.

At this particular juncture in the world's history, and in a work of this kind, it is most necessary to underscore the fact that the author yields to no one in his admiration and respect for Charles Darwin as man and scientist. This essay is in no sense to be regarded as an attack upon Darwin or his work. Nor is anything that is said in these pages to be construed as lending support to that aberration of the human mind known as Lysenkoism. This essay represents no more than an attempt to indicate the kind of correction which, in the light of three generations of accumulated knowledge since Darwin wrote, must be made of some of the errors of one of the greatest scientists of all time.

The author trusts that this essay will be sufficiently challenging to stimulate the reader to pursue the inquiry further for himself.

I owe many thanks to Professor Pitirim Sorokin and the Harvard Research Center in Altruistic Integration and Creativity for generous support of the project of which the present work is a part.

In conclusion, the author expresses his thanks to his friends and colleagues Professors Th. Dobzhansky of the Department of Zoology, Columbia University, and Colin Pittendrigh of the Department of Biology, Princeton University, for their critical reading of the manuscript and their helpful suggestions.

ASHLEY MONTAGU

Department of Anthropology,
Rutgers University,
New Brunswick, New Jersey.

Darwin

COMPETITION
&
COOPERATION

1

DARWIN RE-FOCUSED

THE IMPACT OF DARWINISM ON MANKIND

On the Origin of Species by Means of Natural Selection, Or the Preservation of Favoured Races in the Struggle for Life, by Charles Darwin, was published in London by John Murray on November 24, 1859. Darwin was then fifty years of age. The edition of 1250 copies was sold out on the day of publication. Perhaps no book in the whole history of civilization has made so immediate and enduring an impact upon the world of thought and action as *The Origin of Species*—with the exception, possibly, of the Bible. Darwin's book sought to explain "the mystery of mysteries": the origin—if not of life itself—of species and the diversity of life upon this planet. Similar attempts had been made innumerable times before Darwin, and some writers had even briefly offered the same explanation as he, but Darwin was the first to provide the evidence in support

of the argument—and that is what he called his book [1]
—that this diversity has largely come about, as he says
in the final paragraph of his book, as a result of "a
Ratio of increase [in all forms of life] so high as to lead
to a Struggle for Life, and as a consequence to Natural
Selection, entailing Divergence of Characters and the
Extinction of less-improved forms. Thus, from the war
of nature, from famine and death, the most exalted
object which we are capable of conceiving, namely,
the production of higher animals, directly follows." [2]
Darwin concludes his epochal book with the immediately following words: "There is a grandeur in this
view of life, with its several powers, having been
originally breathed into a few forms or into one; and
that, whilst this planet has gone cycling on according
to the fixed law of gravity, from so simple a beginning
endless forms most beautiful and most wonderful
have been, and are being, evolved."

This is beautifully expressed. But note what Darwin conceives to be the "grandeur" in this view of
life: "The Struggle for Life," "Natural Selection,"
"Extinction," "the war of nature," "famine and death."
It is from these conditions that, according to Darwin,
"the most exalted object which we are capable of conceiving, namely, the production of the higher animals,
directly follows."

This is the view of the nature of evolution that has

---

[1] "As this whole volume is one long argument . . ." *The
Origin of Species,* John Murray, London, 1859, p. 459; 6th
ed., 1872, p. 379.

[2] *Ibid.,* 1859, p. 490; 6th ed., 1872, p. 403.

prevailed from the date of the publication of Darwin's great work to the present day. It is a view that has penetrated and influenced the character of almost every branch of thought and action, and it is a view that has won the support of a great number of the most prominent thinkers of the last ninety years. It is a view that has helped to determine not only the training and conduct of individuals, but also the training and conduct of whole nations. It is a view that has helped to shape relations between persons and between nations in a manner that has led to the most disastrous consequences for whole societies and whole peoples, for millions of unfortunate human beings.

According to this view, life is a struggle for existence in which only the fit survive, the fittest being those who have whatever it takes to survive. Hence, it is argued, those who are successful in exploiting their opportunities, even when those "opportunities" are men or human causes, are thereby biologically justified. Industrialists may exploit and oppress their workers, governments their people; and nations are justified in expropriating the lands of "inferior peoples," for in the struggle for existence natural selection, according to this view, delivers the only just verdict as to who shall come out on top. The "superior" are thus "justified" in suppressing or exterminating ("eliminating" was the more genteel word) the "inferior," and if the poor starve and die it is but nature's decree. War is a good thing, and we shall always have it with us—it is nature's "pruning hook"; as the pruning of trees is calculated to induce a healthy growth, so war acts upon nations to keep them from going to seed, to keep them healthy and strong.[3]

This is the view of the nature of life and of human relations which we have inherited from the nineteenth century, and because it is a view so closely associated with Darwin's work and name, and because it is so demonstrably false, it may be called the Darwinian fallacy. What is solid in Darwin's work will endure. Nothing that anyone may say or wish to say can detract from the greatness of what is sound in Darwin's achievement. There are, however, certain aspects of Darwin's thought and work that are unsound and some that are only partly true. It is these aspects of Darwin's achievement, which have had such disastrous consequences for mankind, with which I am concerned in the pages which follow.

A PHILOSOPHY IN STEP WITH THE TIMES

At the outset it is necessary to emphasize the fact that these consequences did not follow as a direct and exclusive result of Darwin's demonstration of the nature of the evolutionary process. What is now quite clear is that Darwin's conception of that process perfectly fitted the pattern of Victorian social thought and practice. What Darwin's view of life did was to give that social thought and practice a biological validation, a scientific foundation which it had hitherto lacked. Darwin's views were published at a time

---

[3] For some representative statements of these views the reader should turn to Appendix I, pp. 107-113.

when the Industrial Revolution was at its height in England, and in other lands, including America, was well under way. The galloping Imperialism of England, the colonial aspirations of France and the Netherlands, the growing nationalistic spirit of Germany, and the burgeoning industrial enterprise of the United States, had already developed in each of these lands a view of the nature of life, of human relations, and of industry and government, which but awaited the explicit formulation of some acceptable philosopher.

Darwin was born in 1809. He grew to manhood in a period during which wars seemed to be the natural concomitant of living. Darwin was born while Britain was at war against the French in Portugal and Spain (1808-1814), he was six years of age when the British defeated the French at Waterloo (1815), and ten when they took Singapore. Indeed, all through Darwin's life the nations of Europe were frequently engaged in war. During the period just previous to the composition of *The Origin of Species* Britain was engaged in the Crimean War (1854-1855), and the war with Persia and China (1856). In 1857 the Indian "Mutiny" broke out, leading to the frightful massacre at Cawnpore, the "Mutiny" not being suppressed till a year later. Darwin was himself the mildest, the kindest, and the gentlest of men, but he grew up in an age during which war—violent suppression and exploitation of masses of human beings both at home and abroad, by his own people, and what is more by the class to which he belonged—was the order of the day. It was an order in which life for the masses of the people of England was "nasty, brutish, and short." It was as John and Barbara Hammond have termed it in a notable book, *The Bleak Age.*[4] How bleak it was some of us, by far remove, will have learned from the novels of such authors as Charles Dickens and Charles Reade, some from the writings of Robert Owen (1771-1858) and other co-operators, and some from such social histories of the time as we may have read. The human and social degradation of the mass of the people of Darwin's day was unbelievable. The "satanic mills" of the Industrial Revolution had ground human beings into masses, a degraded, starving, struggling flotsam out of which a permanent supply of work could be cheaply siphoned off. Part of the story may be read in the Hammonds' book and in Karl Polanyi's remarkable work *The Great Transformation.*[5]

Darwin himself belonged to the *rentier* class, the upper middle class. In the social struggle for survival he belonged to the class which, by definition, was the fittest. The poor belonged, as Malthus had been at such pains to point out, where they found themselves: to the station which they occupied, to which by God and by Nature they had been called.

The discovery of the real importance of property occurred in the period of Darwin's early life, and as we shall see, it was to play the most important role in the development of Darwin's evolutionary ideas.

---

[4] J. L. and B. Hammond, *The Bleak Age,* Pelican Books, London and New York, 1947.

[5] Rinehart, New York, 1944.

In England this discovery was announced in a work published in 1786, entitled *A Dissertation on the Poor Laws*, by "A Well-Wisher to Mankind." The "Well-Wisher" was a Wiltshire clergyman and noted traveler, the Reverend Joseph Townsend (1739-1816).[6] This important work, so far as I know, has been largely overlooked by writers on the history of the concept of evolution. Townsend illustrated his basic point by recounting what happened on Robinson Crusoe's island, Juan Fernández, off the coast of Chile. A few goats had been landed on the island by the Spanish admiral Juan Fernández (c. 1536-1602?). These had increased at a Biblical rate, providing a convenient store of food for the British privateers and others who were molesting Spanish trade. By landing a greyhound dog and a bitch on the island the Spaniards hoped in a short time to eliminate their enemies by eliminating their food supply. The dogs, so the story goes, multiplied and greatly diminished the number of goats. "Had they been totally destroyed," writes Townsend, "the dogs likewise must have perished. But as many of the goats retired to the craggy rocks, where the dogs could never follow them, descending only for short intervals to feed with fear and circumspection in the vallies, few of these, besides the careless and the rash, became a prey; and none but the most watchful, strong and active of the dogs

---

[6] Joseph Townsend was the author, among other works, of the remarkable *Journey Through Spain*, London, 1791; 3d ed., 1814. Malthus said of this work that it made a fresh examination of Spain unnecessary. Polanyi incorrectly gives Townsend's name as "Wm. Townsend."

could get a sufficiency of food. Thus a new kind of balance was established. The weakest of both species were among the first to pay the debt of nature; the most active and vigorous preserved their lives. It is the quantity of food which regulates the numbers of the human species." Here, succinctly stated, is the principle of natural selection, seventy-three years before the publication of *The Origin of Species,* and twenty-three years before the birth of Darwin.[7] And this, essentially, is the doctrine of Malthus expressed earlier and more clearly by Townsend.

The moral of Townsend's theorem of the dogs and goats is set forth by him in the following maxims which he wished to apply to the reform of the poor laws. But for these maxims, Polanyi believes that "neither Darwin's theory of natural selection, nor Malthus' population laws might have exerted any appreciable influence on modern society." *Op. cit.,* p. 113. Townsend writes, "Hunger will tame the fiercest animals, it will teach decency and civility, obedience and subjection, to the most perverse. In general it is only hunger which can spur and goad them

---

[7] Darwin's statement of the doctrine is as follows: "Hence, as more individuals are produced than can possibly survive, there must in every case be a struggle for existence, either one individual with another of the same species, or with the individuals of distinct species, or with the physical conditions of life. It is the doctrine of Malthus applied with manifold force to the whole animal and vegetable kingdoms." Charles Darwin, *The Origin of Species,* London, John Murray, 1859, Chapter 3, p. 63.

[the poor] on to labour; yet our laws have said they shall never hunger. The laws, it must be confessed, have likewise said, they shall be compelled to work. But then legal constraint is attended with much trouble, violence and noise; creates ill will, and never can be productive of good and acceptable service; whereas hunger is not only peaceable, silent, unremitting pressure, but, as the most natural motive to industry and labour, it calls forth the most powerful exertions; and, when satisfied by the free bounty of another, lays lasting and sure foundations for good will and gratitude. The slave must be compelled to work but the free man should be left to his own judgment, and discretion; should be protected in the full enjoyment of his own, be it much or little; and punished when he invades his neighbour's property." Thus, with the assistance of dogs and goats, were the laws of nature introduced into political science, and hunger made the deliberate and principal means of good government!

MALTHUS PROVIDES THE THEORY

The story of the goats and dogs is not wholly true. Fernández did land goats, but dogs were not introduced, only cats. Neither goats nor cats are known to have multiplied. In any event, the significance of this tale does not depend upon its historical veracity. Its significance lies in the fact that it gave a theory to and influenced the thought of perhaps the most influential writer on population studies who ever put pen to paper, namely, Thomas Robert Malthus (1766-1834), and through him provided Charles Darwin with the theory of evolution. Townsend's goats and dogs maintained a balance by the difficulty which the dogs had in following the goats into the rocky part of the island, and the inconveniences the goats had to face when moving to safety from the dogs. The fleet-footed dogs got the slow-footed goats, and both the goats and the dogs starved and died. Whatever increase of each there might be, the balance was always restored by the pangs of hunger on the one hand and the scarcity of food on the other. Men, claimed Townsend, were no different than goats and dogs, insofar as the balance of nature was concerned. A free society was made up of two races, property owners and laborers, and the number of the latter was limited by the amount of food.

Malthus published his *Essay on Population* anonymously in 1798,[8] and a second edition in 1803. In this *Essay* Malthus points out that "the struggle for existence"—this is his own phrase—is an ever-present fact of life. The members of any group of animals are simply the fittest—those who have survived the struggle for existence. The principle of population is in the foreground; and there are no checks to it but famine,

---

[8] [Thomas R. Malthus], *An Essay on the Principle of Population as It Affects the Future Improvement of Society*, J. Johnson, London, 1798. For an excellent discussion of the history of this subject see Kenneth Smith, *The Malthusian Controversy*, Routledge and Kegan Paul, London, 1951.

disease, and death. Unlike Condorcet (1743-1794)[9] and Godwin (1756-1836),[10] Malthus saw an insurmountable obstacle in the way of man's infinite progress. How is progress possible, asks Malthus—who in addition to being a clergyman was a mathematician—when population increases in geometrical ratio, while the means of subsistence increases only in arithmetical ratio? (. . . that is, as 1, 2, 4, 8, 16, 32, 64, 128, 256, as compared with 1, 2, 3, 4, 5, 6, 7, 8, 9). "A slight acquaintance with numbers," Malthus drily observes, "will show the immensity of the first power in comparison with the second." He goes on to show that "the race of plants and animals shrinks under the great restrictive law, and the race of men cannot by any efforts of reason escape from it. Among plants and animals its effects are waste of seed, sickness, and premature death; among men, misery and vice."

It was the reading of Malthus' *Essay* in October, 1838, that gave Darwin the theory for which he had been seeking. "I happened," he tells us, "to read for amusement 'Malthus on Population,' and being well prepared to appreciate the struggle for existence which everywhere goes on from long-continued observation of the habits of animals and plants, it at once struck me that under these circumstances fa-

---

[9] Marquis de Condorcet (1743-1794), French mathematician, philosopher, and political leader. See his *Outlines of an Historical View of the Progress of the Human Mind*, London, 1795.

[10] William Godwin, English author and political philosopher; see *Political Justice*, London, 1792, and *Of Population*, London, 1820.

vourable variations would tend to be preserved, and unfavourable ones to be destroyed. The result of this would be the formation of new species. Here then I had at last got a theory by which to work." [11] Some twenty years later after reading Malthus the same ideas occurred to Alfred Russel Wallace (1823-1913), and it was his communication of these ideas to Darwin that, at the urging of Wallace and other friends, caused Darwin to write and publish what he called an "Abstract," in the form of *The Origin of Species*, of the much larger work which he hoped, but was never able, to produce.

DARWIN EXTENDS THE THEORY

In *The Origin of Species* Darwin acknowledges his debt to Malthus in his "Introduction." He writes, "In the next chapter [Three] the Struggle for Existence [Malthus' phrase] amongst all organic beings throughout the world, which inevitably follows from their high geometrical powers of increase ['geometrical ratio of increase' in later editions], will be treated of. This is the doctrine of Malthus, applied to the whole animal and vegetable kingdoms. As many more individuals of each species are born than can possibly survive; and as, consequently, there is a frequently recurring struggle for existence, it follows that any

---

[11] Charles Darwin, "Autobiography," in *The Life and Letters of Charles Darwin* (edited by Francis Darwin), John Murray, London, 1888, vol. 1, p. 83; *Charles Darwin's Autobiography* (edited by Francis Darwin), Schuman, New York, 1950, p. 54.

being, if it vary however slightly in any manner prof-
itable to itself, under the complex and sometimes
varying conditions of life, will have a better chance
of surviving, and thus be *naturally selected*. From the
strong principle of inheritance, any selected variety
will tend to propagate its new and modified form." [12]
And this is what Darwin seeks to prove in his book.
It is, as Darwin states, "the doctrine of Malthus,
applied to the whole animal and vegetable kingdoms." Malthus did not pursue his generalization beyond man. Darwin applied it to the whole kingdom
of life. As one writer has put it, "Darwinism is Mal-
thusianism on the largest scale." [13]

What I have thus far attempted to indicate is that
Darwin's explanation of the origin of species by
means of natural selection, or the preservation of

---

[12] *The Origin of Species*, 1859, pp. 4-5; 6th ed., 1872,
p. 3. I reproduce here one of the clearer of the nineteenth
century statements of the Darwinian conception of natural selection. It is by Thomas Henry Huxley. "All *species*
have been produced by the development of *varieties* from
common stocks: by the conversion of these, first into
permanent races and then into *new species*, by the process of *natural selection*, which process is essentially identical with that artificial selection by which man has
originated the races of domestic animals—the *struggle for
existence* taking the place of man, and exerting, in the
case of natural selection, that selective action which he
performs in artificial selection." "Darwiniana," *Collected
Essays*, vol. 2, 1893, p. 71.

[13] Robert M. Murray, *Studies in the English Social and
Political Thinkers of the Nineteenth Century*, Heffer,
Cambridge (England), 1929, vol. 1, p. 6.

favored races in the struggle for life, largely owes its character to social and political thinkers of the late eighteenth and early nineteenth centuries.[14] Darwin was no less a child of his time than we are of ours. As an upper middle class gentleman he absorbed the values of his period, and in seeking for a theory to explain his accumulated facts Darwin found that Malthus' principle of population fitted both the values and the facts like a glove. The human struggle was paralleled by the animal struggle. It was not that the human struggle was seen as a part of the struggle of nature, but rather that nature was interpreted in terms of the struggle for existence of men living or attempting to live in a ruthless industrial society in which the fittest alone survived. The fittest inherited what it took to survive,[15] and by safeguarding and augmenting what they had, the favored classes were

---

[14] Judd states that "The idea that as animals and plants multiply in geometrical progression, while the supplies of food and space to be occupied remain nearly constant, and that this must lead to a 'struggle for existence' of the most desperate kind, was by no means new to Darwin, for the elder De Candolle, Lyell and others had enlarged upon it; yet the facts with regard to the human race, so strikingly presented by Malthus, brought the whole question with such vividness before him that the idea of 'Natural Selection' flashed upon Darwin's mind." J. W. Judd, *The Coming of Evolution*, Cambridge University Press, New York, 1925, p. 107.

[15] "The greater or less force of inheritance and reversion determine whether variations shall endure." *The Origin of Species*, 1859, p. 43; 6th ed., 1872, p. 30.

preserved in the struggle for existence, whereas those who inherited no property and were forced to work for a living would leave nothing to transmit to their progeny but their sins. This, in fact, is a brief description of the pattern of the structure of social thought in England during Darwin's day.[16] It largely influenced Malthus' thought, and it substantially determined the manner in which Darwin was ready to perceive the meaning of his accumulated facts. What is perceived is largely preconceived. It is no mere coincidence that two naturalists, Darwin and Wallace, independently reading Malthus, should be struck by the same idea, namely, the application of the Malthusian principle to the explanation of the origin of species.

I believe it was the late Professor Patrick Geddes (1854-1932) who was the first to draw attention to the fact that Darwin's central idea was correlated with the state of contemporary social evolution. "The

---

[16] This pattern of thought had a much longer history, having already found a happy medium for development among the Puritans of seventeenth century England. As Knowles puts it, "Many of them considered that to be poor was to prove that the Lord had turned His countenance away; those whom the Lord had blessed prospered. The incentive to 'get on' was very strong when the acquisition of wealth seemed to be an earnest of ease and comfort both here and hereafter." L. C. A. Knowles, *The Industrial and Commercial Revolutions in Great Britain During the Nineteenth Century*, Routledge, London, 1947, p. 174.

substitution of Darwin for Paley," [17] wrote Geddes, "as the chief interpreter of the order of nature is currently regarded as the displacement of an anthropomorphic view by a purely scientific one: a little reflection, however, will show that what has actually happened has been merely the replacement of the anthropomorphism of the eighteenth century by that of the nineteenth. For the place vacated by Paley's theological and metaphysical explanation has simply been occupied by that suggested to Darwin and Wallace by Malthus in terms of the prevalent severity of industrial competition, and these phenomena of the struggle for existence which the light of contemporary economic theory has enabled us to discern, have thus come to be temporarily exalted into a complete explanation of organic progress." [18] These words were written in 1882.

The validity of a theory is unaffected by what suggested it, but it does help us better to understand something of its nature when we know the conditions

---

[17] Archdeacon William Paley (1743-1805), A View of the Evidences of Christianity, 1794, and especially his Natural Theology, or Evidences of the Existence and Attributes of the Deity Collected From the Appearances of Nature, 1802, exerted a powerful influence upon the thought of English-speaking peoples throughout the greater part of the nineteenth century. In the spirit of rationalism Paley attempted to prove the truth of Christian doctrine.

[18] Patrick Geddes, "Biology," in Chambers's Encyclopaedia, Edinburgh, London, and Philadelphia, 1882, vol. 4, p. 116.

that led to its origin. In a machine age Darwin gave a mechanical explanation of evolution. In an age characterized by industrial competition in which no quarter was given, Darwin gave an explanation virtually entirely in terms of competition,[19] in terms of the struggle for life or existence, the survival of the fittest. Darwin, in short, gave an explanation of the origin of species in terms of the spirit of his time.[20] The contemporary background against which the development of the Darwinian theory must be viewed is the suffocating atmosphere of overpopulation in nineteenth century industrializing and urbanizing England.

---

[19] "As natural selection acts by competition, it adapts and improves the inhabitants of each country only in relation to the co-inhabitants." *The Origin of Species*, 6th ed., 1872, p. 389. In the first edition (1859, p. 471) this reads, "As natural selection acts solely by accumulating slight, successive, favourable variations, it can produce no great or sudden modifications."

[20] Karl Marx (1818-1883) was so delighted with Darwin's demonstration of nature's warfare, since it gave him a biological foundation for his conception of the warfare of the classes, that he thought of dedicating *Das Kapital* (vol. 1, 1867) to Darwin. Reading *The Origin of Species* in 1860, Marx reported to Engels and later declared to LaSalle, that "Darwin's book is very important and serves me as a basis in natural science for the class struggle in history." *The Correspondence of Marx and Engels*, International Publishers, New York, 1935, pp. 125-126. See also Jacques Barzun, *Darwin, Marx, Wagner*, Little, Brown, Boston, 1941.

SPENCER DEVELOPS "SOCIAL DARWINISM"

I want to make this point clear because it is often assumed that social thought after 1859 was largely the social reflection of Darwin's biology. The truth is that Darwinian biology was largely influenced by the social and political thought of the first half of the nineteenth century, and that its own influence took the form of giving scientific support in terms of natural law for what had hitherto been factitiously imposed social law.[21] Darwin unwittingly provided the nineteenth century with a philosophy of industrial progress. He provided the age with its supreme rationalization—a rationalization, however, with full-blown scientific support. The philosophy of industrial progress that had been so conveniently provided by Darwin was actually developed by Herbert Spencer (1820-1903).[22] What Darwin provided was a scientific description of how evolution worked in nature. What Spencer did was to describe, as he thought, how nature worked in society, a doctrine which came to be called "Social Darwinism." Lock, stock, and barrel,

---

[21] The distinguished prehistorian, Professor V. Gordon Childe, has recently made this point. He writes, "it seems that Darwin's contemporaries applied as an analogy to organic nature the prevailing (but erroneous) conception of economic order and progress being the product of *laissez-faire* regime of unrestricted competition." *History,* Cobbett Press, London, 1947, p. 55.

[22] Herbert Spencer, English philosopher and sociologist, in social science the chief exponent of the doctrine of "Social Darwinism."

Spencer applied the concepts developed by Darwin to the interpretation of the nature and functioning of society. What was thought to apply to the physical organism was transferred to the social organism, and the alleged universal competition of organisms in a state of nature was extrapolated to apply to man in society. Darwin and Spencer gave the industrialists of the nineteenth century a cosmic sanction for free competition: "Just as in their primitive struggle for existence the 'fittest' among the species of sea and forest adapted themselves to their environment, so for Spencer those competitors who had best adapted themselves to nineteenth century society, became the 'fittest' among men. And just as nature worked untrammeled in 'selecting' her elite, so that society was headed quickest to perfection which allowed its elite free play." [23]

EARLY CRITICS OF DARWIN'S CONCEPTION

In the years since the publication of *The Origin of Species* voices which urged that the Darwinian con-

---

[23] Thomas C. Cochran and William Miller, *The Age of Enterprise,* Macmillan, New York, 1942, p. 122. In this and in the following three works will be found excellent accounts of Social Darwinism and its influence: George Nasmyth, *Social Progress and the Darwinian Theory,* Putnam, New York, 1916; Richard Hofstadter, *Social Darwinism in American Thought, 1860-1915,* University of Pennsylvania Press, Philadelphia, 1944; Stow Persons (editor), *Evolutionary Thought in America,* Yale University Press, New Haven, Conn., 1950.

ception of evolution was one-sided and incomplete
have not been wanting. I do not refer to the ignorant,
malicious, and prejudiced attacks which, ever since
its announcement, have been made upon the theory
of evolution by those having vested interests in special theories or systems of belief. Nor do I refer to
those attacks and criticisms which may most chari-
tably be described as indefensible and uninformed.
I have in mind the scientists and scientific thinkers
in Darwin's own and allied fields who, though having
the respect and admiration for Darwin which he so
fully merits, have endeavored to show that he and
such followers of his as Herbert Spencer and Thomas
Henry Huxley (1825-1895) [24] overstressed and mis-
understood the factor of competition and either
largely or altogether neglected and underestimated
the factor of co-operation. They failed to give the
factor of co-operation a place in the concept of natural selection in particular and in the description of
the evolutionary process in general. In this substan-
tive failure to recognize the importance of co-operation they succeeded in conveying a view of nature
that was badly maimed and sadly out of focus.
Among those who attempted to present a more
rounded view of evolution may be mentioned Espinas in France in 1878, Kessler in Russia in 1880, Lanessan in France in 1882, Büchner in Germany in
1883, Kropotkin in England in 1888, Geddes and
Thomson in England in 1889, and Drummond in

---

[24] Thomas Henry Huxley, English biologist; "Darwin's
bulldog," as he described himself, and in general Darwin's apostle to the Gentiles.

England and the United States in 1894,[25] to name
some of the outstanding critical voices of the nine-

---

[25] Much valuable material on the importance of co-
operation in evolution by writers of the twentieth century
will be found in the following: C. Bouglé, "Darwinism
and Sociology," in *Darwin and Modern Science* (edited
by A. C. Seward), Cambridge University Press, New
York, 1909, pp. 465-476; Henry M. Bernard, *Some Neglected Factors in Evolution*, Putnam, New York, 1911;
Patrick Geddes and J. Arthur Thomson, *Sex*, Williams &
Norgate, London, 1911; Yves Delage and Marie Goldsmith, *The Theories of Evolution*, Huebsch, New York,
1912; Hermann Reinheimer, *Evolution by Co-operation:
A Study of Bioeconomics*, Kegan Paul, London, 1913;
George Nasmyth, *Social Progress and the Darwinian The-
ory*, Putnam, New York, 1916; P. Deegener, *Die Formen
der Vergesellchaftung im Tier-reiche*, Veit, Leipzig, 1918;
John M. Macfarlane, *The Causes and Course of Organic
Evolution*, Macmillan, New York, 1918; William Patten,
The Grand Strategy of Evolution, Badger, Boston, 1920;
Hermann Reinheimer, *Symbiosis: A Socio-Physiological
Study of Evolution*, Headley, London, 1920; Robert W.
Gibson, *The Morality of Nature*, Putnam, New York,
1923; Leo S. Berg, *Nomogenesis, or Evolution Determined
by Law*, Constable, London, 1926; William M. Wheeler,
Social Life Among Insects, Harcourt, New York, 1923;
E. Wales Hirst, *Ethical Love*, Allen & Unwin, London,
1928; William M. Wheeler, "Societal Evolution," in *Hu-
man Biology and Racial Welfare* (edited by Edmund V.
Cowdry), Hoeber, New York, 1930; Herbert F. Standing,
Spirit in Evolution, Allen & Unwin, London, 1930; Warder
C. Allee, *Animal Aggregations*, University of Chicago
Press, Chicago, 1931; Warder C. Allee, *The Social Life of*

Animals, 1st ed., Norton, New York, 1938, 2d ed., *Cooperation Among Animals,* Schuman, New York, 1951; Christopher Caudwell, "Love," in *Studies in a Dying Culture,* Bodley Head, London, 1938; William Galt, "The Principle of Co-operation in Behavior," *Quarterly Review of Biology,* vol. 15, 1940, pp. 401-410; Charles Sherrington, *Man on His Nature,* Cambridge University Press, New York, 1941; Robert Redfield (editor), "Levels of Integration in Biological and Social Systems," *Biological Symposia,* vol. 8, Jaques Cattell Press, Lancaster, Pa., 1942; Alfred E. Emerson, "Basic Comparisons of Human and Insect Societies," in *Biological Symposia,* vol. 8, 1942, pp. 163-177; R. Gerard, "Higher Levels of Integration," in *Biological Symposia,* vol. 8, 1942, pp. 67-78; Ralph S. Lillie, *General Biology and Philosophy of Organism,* University of Chicago Press, Chicago, 1945; Alfred E. Emerson, "The Biological Basis of Social Cooperation," *Transactions, Illinois Academy of Science,* vol. 39, 1946, pp. 9-18; L. R. Wheeler, *Harmony of Nature: A Study of Cooperation for Existence,* Longmans, New York, 1947; Thomas H. Huxley and Julian S. Huxley, *Touchstone for Ethics,* Harper, New York, 1947; F. J. Trembley, "Evolution and Human Affairs," *Proceedings of the Pennsylvania Academy of Science,* vol. 23, 1949, pp. 181-195; M. F. Ashley Montagu, "The Origin and Nature of Social Life and the Biological Basis of Coöperation," *Journal of Social Psychology,* vol. 29, 1949, pp. 267-283; Samuel J. Holmes, *Life and Morals,* Macmillan, New York, 1948; Herman J. Muller, "Genetics in the Scheme of Things," *Hereditas,* Supplementary Volume, 1949, pp. 96-127; Hugh Miller, *The Community of Man,* Macmillan, New York, 1949; E. Morton Miller, "A Look at the Anatomy and Physiology of Groups," *Bios,* vol. 20, 1949, pp. 24-31; Warder C. Allee *et al., Principles of Animal Ecology,*

teenth century.[26] With the exception of Kropotkin and Drummond, all these writers were professional biologists. Kropotkin and Drummond had received a scientific training. Of all these authors Kropotkin

---

Saunders, Philadelphia, 1949; George C. Simpson, *The Meaning of Evolution*, Yale University Press, New Haven, Conn., 1949; Ashley Montagu, *On Being Human*, Schuman, New York, 1950; Pitirim A. Sorokin (editor), *Explorations in Altruistic Love and Behavior*, Beacon, Boston, 1950.

[26] See also A. V. Espinas, *Des Sociétés animales*, Librairie Baillière, Paris, 1878, 3d ed., 1924; Professor Kessler, "Mutual Aid as a Law of Nature and the Chief Factor of Evolution," *Memoirs (Trudy) of the St. Petersburg Society of Naturalists*, vol. 9, a lecture delivered at the annual meeting of the St. Petersburg Society of Naturalists, January 8, 1880, noticed in *Nature* (London), January 21, 1880; Ludwig Büchner, *Aus dem Geistesleben der Thiere*, 2d ed., 1877, and the same author's *Liebe und Liebesleben in der Thierwelt*, Berlin, 1879; J. M. A. Lanessan, "La Lutte pour l'Existence et l'Association pour la Lutte," *Bibliothèque Biologique*, Paris, 1881; Patrick Geddes and J. Arthur Thomson, *The Evolution of Sex*, Scott, London, 1889, Scribner & Welford, New York, 1890; Menzbir, *Darwinism in Biology* (in Russian). See also such works as Henry Drummond, *The Ascent of Man*, Hodder & Stoughton, London, 1894; Henry George, *Progress and Poverty*, New York, 1879 (especially Book X, Chapter III); Edward Bellamy, *Looking Backward*, New York, 1888; J. Novicow, *Les Luttes entre Sociétés humaines et leur Phases successives*, Paris, 1893, 2d ed., 1896; see also Novicow's *la Guerre et ses Prétendus Bienfaits*, Paris, 1894.

(1842-1921), I believe, has made the most enduring impression.

HUXLEY VS. KROPOTKIN—COMPETITION VS. CO-OPERATION

In February, 1888, T. H. Huxley published his famous "struggle for life" manifesto entitled "The Struggle for Existence: A Programme." [27] In this article Huxley declared that "from the point of view of the moralist, the animal world is on about the same level as a gladiator's show. The creatures are fairly well treated, and set to fight—whereby the strongest, the swiftest and the cunningest live to fight another day. The spectator has no need to turn his thumbs down, as no quarter is given." And as among animals, so among primitive men "the weakest and stupidest went to the wall, while the toughest and shrewdest, those who were best fitted to cope with their circumstances, but not the best in any other sense, survived. Life was a continual free fight, and beyond the limited and temporary relations of the family, the Hobbesian war of each against all was the normal state of existence. The human species, like others, plashed and foundered amid the general stream of evolution, keep-

---

[27] In the periodical *The Nineteenth Century* (London), vol. 23, February, 1888, pp. 161-180, reprinted in Huxley's *Collected Essays*, vol. 5. The "manifesto" provided the main thesis for Huxley's Romanes Lecture, "Evolution and Ethics," delivered at Oxford May 18, 1893, for which see T. H. and J. S. Huxley, *Touchstone for Ethics*, Harper, New York, 1947.

ing its head above water as it best might, and thinking neither of whence nor whither." [28]

Actually, the struggle for existence with which Huxley is chiefly concerned in this article is the struggle of an overpopulated industrial England in an highly competitive, commercialized world. Huxley's "Programme" offers a solution in terms of raising the standards of living and better technical education of the masses.

Kropotkin was convinced that Huxley had given "a very incorrect representation of the facts of Nature, as one sees them in the bush and in the forest," and between September, 1890, and June, 1896, in a series of eight articles, replied to Huxley's "gladiatorial" view of evolution by setting out the facts for animals and man. Huxley almost certainly read most of Kropotkin's articles, but if he did he made no direct printed reference to them, though it is quite possible that their influence is to be seen in Huxley's Romanes lecture of 1893. In 1902 Kropotkin's articles were published in book form with the title *Mutual Aid: A Factor of Evolution*. Kropotkin showed, that—in his own words—"Happily enough, competition is not the

---

[28] *The Nineteenth Century* (London), vol. 23, February, 1888, p. 163 and p. 165. For an interesting review of this paper see "Messrs Goschen and Huxley on English Culture," *Nature* (London), vol. 37, February 9, 1888, pp. 337-338. The editors write, "This application of Darwin's great theory to commercial competition is more than a parallel. It is the scientific application of causes which have wrecked civilizations in the past and may wreck them in the future."

rule either in the animal world or in mankind. It is
limited among animals to exceptional periods, and
natural selection finds better fields for its activity.
Better conditions are created by the *elimination of
competition* by means of mutual aid and mutual
support.[29] In the great struggle for life—for the greatest possible fulness and intensity of life with the least
waste of energy—natural selection continually seeks
out the ways precisely for avoiding competition as
much as possible. The ants combine in nests and nations; they pile up their stores, they rear their cattle
—and thus avoid competition; and natural selection
picks out of the ants' family the species which know
best how to avoid competition, with its unavoidably
deleterious consequences. Most of our birds slowly
move southwards as the winter comes, or gather in
numberless societies and undertake long journeys—
and thus avoid competition. Many rodents fall asleep
when the time comes that competition should set
in; while other rodents store food for the winter,
and gather in large villages for obtaining the necessary protection when at work. The reindeer, when
the lichens are dry in the interior of the continent,
migrate towards the sea. Buffaloes cross an immense
continent in order to find plenty of food. And the
beavers, when they grow numerous on a river, divide

---

[29] [Kropotkin's note.] " 'One of the most frequent modes
in which Natural Selection acts is, by adapting some individuals of a species to a somewhat different mode of life,
whereby they are able to seize unappropriated places in
Nature' (*Origin of Species*, p. 145)—in other words, to
avoid competition."

into two parties, and go, the old ones down the river, and the young ones up the river—and avoid competition. And when animals can neither fall asleep, nor migrate, nor lay in stores, nor themselves grow their food like the ants, they do what the titmouse does, and what Wallace (*Darwinism,* ch. v) has so charmingly described: they resort to new kinds of food— and thus, again, avoid competition.

" 'Don't compete!—competition is always injurious to the species, and you have plenty of resources to avoid it!' That is the *tendency* of nature, not always realized in full, but always present. That is the watchword which comes to us from the bush, the forest, the river, the ocean. 'Therefore combine—practice mutual aid! That is the surest means for giving to each and to all the greatest safety, the best guarantee of existence and progress, bodily, intellectual, and moral.' That is what Nature teaches us." [30]

The conception of competition which Kropotkin is here attacking is the "nature, red in tooth and claw" view of it, the "gladiatorial" view of "the struggle for existence" expounded by such interpreters of Darwin as Thomas Huxley. The fact is that Kropotkin had a much more accurate conception of the nature of competition as a process of evolution than most nineteenth century biologists. Indeed, his whole book is devoted to demonstrating that competition is not necessarily the brutally ruthless process that most Darwinians conceived it to be, but that, in his own

---

[30] Petr Kropotkin, *Mutual Aid: A Factor of Evolution,* Penguin Books, London, 1939, pp. 72-73.

words, "In the great struggle for life" the struggle is "for the greatest possible fulness and intensity of life with the least waste of energy."

It cannot be too emphatically stated that Kropotkin did not consider that the demonstrable importance of co-operation or mutual aid as *a* factor of evolution in any way contradicted the theory of natural selection. What he sought to show was that the theory of natural selection is incomplete if it neglects such an important factor as mutual aid, and that that theory does violence not only to the facts but to the possibility of man's own social evolution. Kropotkin did not call his book "Mutual Aid: *The* Factor of Evolution," but he called it "Mutual Aid: *A* Factor of Evolution." As the title of his book implies and as his text seeks to show, Kropotkin considered mutual aid to be a factor of evolution, not *the* only factor.

Kropotkin's book is now a classic—which means that few people read it and that it is probably out of print. Yet no book in the whole realm of evolutionary theory is more readable or more important, for it is *Mutual Aid* which provided the first thoroughly documented demonstration of the importance of co-operation as a factor in evolution.[31] Kropotkin's book, one may be sure, is destined for a revival, and the influence it has already had is likely to increase manyfold with the years.

Since the publication in book form of *Mutual Aid*

---

[31] Kropotkin developed the principle of Mutual Aid as applied to human relations in his posthumously published book, *Ethics: Origin and Development,* Tudor, New York, 1947.

an increasing number of books and studies have been published along similar lines.

"COMPETITIVE DARWINISM" GIVES WAY

Today, an increasingly large number of individual biologists are becoming more critical of the Darwinian conception of evolution. I say "individual biologists" deliberately, because the orthodox canon of biological thought yet remains virtually unaltered, namely, that the central doctrine of evolution is the principle that those organisms which are better adapted to their environment "replace through competition the less well-adapted individuals of the species. This is the process Darwin called natural selection, and Spencer called the survival of the fittest." [32]

There can, however, remain little doubt that the canon of "competitive Darwinism," as it may be called, will not much longer be able to withstand the pressure of the facts.[33] One may be certain that were Darwin alive today he would be among the first to accept them.

---

[32] Thomas Hunt Morgan, *Evolution and Genetics,* Princeton University Press, Princeton, N. J., 1932, p. 120.

[33] When the distinguished Harvard biologist William Morton Wheeler (1865-1937) wrote, in 1928, of "the petering out during the past seventy years of the theory of the survival of the fittest," one cannot help but think that his acceleration of the process was more the expression of a wish than a statement of the fact. See Wheeler's "Present Tendencies in Biological Thought," reprinted in his *Essays in Philosophical Biology,* Harvard University Press, Cambridge, Mass., 1939, p. 205.

Kropotkin's remarks on competition seem to have fallen into a bottomless well so far as most biologists are concerned. More than fifty years after Kropotkin's elaborate discussion of it, S. J. Holmes, Professor Emeritus of Zoology at the University of California, apparently quite independently of Kropotkin, arrived at a virtually identical view of the part played by competition in evolution. "Competition," Holmes writes, "makes for diversity because, if I may speak figuratively, life is continually endeavoring to escape from it. In organized society the avenues of escape normally lead to activities requiring mutual aid and co-operation. Societies are mutual benefit associations and they tend to engender an altruism among their members that is quite closely subordinated to the egoistic interests of the group as a biological unit. One great advantage of social life is that it secures the benefits of competition and co-operation at the same time." [34]

The same holds true for the organism. In the organism there is a balanced interrelation between "competitive" and co-operative activities. Competition in this sense means that every component of the organism "strives" to achieve the same end, namely, the satisfactory functioning of the parts mutually involved. In this way the survival and adequate functioning of the organism as a whole is achieved.

[34] Samuel J. Holmes, "The Problem of Organic Form. II: Competition as an Integrative Force," *Scientific Monthly*, vol. 59, 1949, p. 253. See also the same author's *Organic Form*, University of California Press, Berkeley, 1948, pp. 23-28.

DEFINITIONS OF "COMPETITION"

This is the kind of competition that Darwin did *not* have in mind when he used the word. Thus, in *The Origin of Species* he wrote, "I should premise that I use the term Struggle for Existence in a large and metaphorical sense, including dependence of one being upon another, and including (which is more important) not only the life of the individual, but success in leaving progeny." (Chapter III, p. 62.) Having mentioned "dependence," Darwin promptly drops the notion and proceeds to use Struggle for Existence in terms of competition. As Professor Holmes points out, "It may be said also that there is competition between co-operating elements and that this circumstance . . . [makes] for a balancing or automatic adjustment of activities." [35]

Darwin, too, saw competition as a striving to achieve the same end, and though he asserted that he used the word "struggle"—his word for competition—in a metaphorical sense, in actual usage he more often than not gave the word a content and a context of conflict, of combativeness. For Darwin, "competition" or "struggle" was usually ruthless and uninhibited. For those who were at pains to interpret Darwin, "struggle" or "competition" *was* "combat"—in nature, "red in tooth and claw;" in society, the mailed fist or gloved talon. Darwin borrowed the conception and the phrase "the struggle for existence," the idea of "competition," from Malthus. Originally

[35] *Ibid.*, p. 253.

used with reference to the conditions prevailing among human beings in a mercantile civilization, Darwin took over the conception and used it to interpret the relations existing between all living things. What Darwin did, without being fully aware of it, was to take over the notion prevailing in the mercantile world of his day that "the war of commerce, carried on under the name of competition, constitutes the life of trade." Translated into the realm of biology by Darwin this became "the war of nature, carried on as the struggle for existence, is what preserves favoured races."

WAR: AN UNNATURAL PHENOMENON

There is, in fact, no such thing as "the war of nature." War is not a natural but an *unnatural* phenomenon. Indeed, as Leonardo da Vinci (1452-1519) stated long ago, man is the only animal which persecutes its *own* as well as other living species. The "wars" of apocryphal ants and other creatures are purely imaginary. No creature other than man makes war. Man is the only creature which makes organized attacks, war, upon its own species.[36]

The "struggle for existence" in nature has been greatly overdone. It isn't at all like the struggle between nineteenth century business men. Indeed, in

---

[36] See M. F. Ashley Montagu, "The Nature of War and the Myth of Nature," *Scientific Monthly,* vol. 54, 1942, pp. 342-353; M. F. Ashley Montagu, *Man's Most Dangerous Myth: The Fallacy of Race,* 3d ed., Harper, New York, 1952.

the oft-cited tropical "jungle"—allegedly dripping with poisonous snakes and pullulating with predatory panting panthers—conditions, apparently, "are so favorable that almost anything can survive, and almost everything does. Individuals may have a hard time, but some of them seem always to manage to leave offspring; and no one kind of animal or plant is able to get the upper hand of all the rest, and thus dominate the landscape." [37]

DARWIN'S FUNDAMENTAL ERROR

The fact is, the fundamental error committed by Darwin was to take over the Malthusian doctrine—which Malthus had specifically applied to men living in an industrial society—and apply it to the whole vegetable and animal kingdom. What little was biologically sound in the idea as applied to man remains so, but what represented the accretions of the matrix of the competitive predatory society in which the idea was developed, adhered to it in the form of such conceptions as "war," "the struggle for existence," "competition," "famine and death," "the survival of the fittest," and the like.

This is the view of life of men living and conditioned in an industrial society, the way of life of nations of "shopkeepers," *but it is not life.* It is not even the way of life lived by most human beings on the face of the earth today. It is not a way of life that is even remotely approached by any other ani-

[37] Martson Bates, *Where Winter Never Comes,* Scribner, New York, 1952, p. 209.

mal. It is the way of life of men living in industrial civilizations. The rest of the animal world does not live according to the principles prevailing in nineteenth century competitive mercantile civilizations. It would constitute one of the wriest jokes of history (were it not for the fact that it has proven so tragic in its consequences) that the "cosmic process," as T. H. Huxley termed it, should have been envisaged as functioning after the pattern of perhaps the most predatory industrial civilization that the world has ever known. Man sees the world according to the kingdom that is within him; it should, therefore, not be surprising that nature should have been interpreted in terms of nineteenth century human relations.

On the contrary, organisms in themselves and in relation to their fellows in their functions are more akin to societies in which everyone has a part in determining the government as a whole. As Professor Holmes has recently stated, "The survival of the organism must depend primarily upon the aptitude of its members for getting on well together. The groups in which the constituents behave at cross-purposes quickly go into the discard. . . . The self-perpetuating assemblage of genetic factors is mostly a well-ordered body whose members for the most part cooperate most admirably to promote the common weal. Government, as in societies of insects, seems, on the whole, to be on a democratic basis, which, after all, is the organismic method of regulation." [38]

---

[38] S. J. Holmes, "What Is Natural Selection?" *Scientific Monthly*, vol. 67, 1948, pp. 324-330.

THE CONCEPTION OF "DEFINITE VARIATION"

It is more than seventy years since that extraordinary genius, Patrick Geddes (1854-1932), then Professor of Biology in the University of Edinburgh, penned the following words: ". . . it is being attempted to replace the received doctrine of indefinite variation, with progress by means of struggle for existence among individuals, by the conception of definite variation (even pathological), with progress essentially in the measure of the subordination of individual struggle and development to species-maintaining ends, reproductive, domestic, or associative. Without entering into details, it is evident that such a restatement of the theory of the evolution of living beings—in terms no longer primarily of strength and competition, of hunger and battle, but of love and co-operation, of sacrifice and pain—would involve, no less fully than has the doctrine of struggle for existence, a deepened reinterpretation of plant and animal life, and would similarly extend into other fields than those of pure biology." [39]

What Geddes wrote as "being attempted" in 1882 has even yet, more than seventy years later, not got fully under way. In a book, *The Evolution of Sex,* published in collaboration with his fellow-biologist J. Arthur Thomson, the authors pointed out "that

---

[39] Patrick Geddes, article "Biology" in *Chambers's Encyclopaedia,* London, Edinburgh, and Philadelphia, 1882, p. 161. See also Geddes and J. Arthur Thomson, *The Evolution of Sex,* Scott, London, 1889, pp. 279-281, pp. 311-314.

increase of reproductive sacrifice . . . increase of parental care . . . [and] that frequent appearance of sociality or co-operation, which even in its rudest forms so surely secures the success of the species attaining it, be it mammal or bird, insect or even worm,—all these phenomena of survival of the truly fittest, through love, sacrifice, and co-operation, need far other prominence than they could possibly receive on the hypothesis of the essential progress of the species through internecine struggle of its individuals at the margin of subsistence. Each of the greater steps of progress is in fact associated with an increased measure of subordination of individual competition to reproductive or social ends, and of interspecific competition to co-operative association.

"The corresponding progress in the historic and individual world, from sex and family up to tribe or city, nation and race, and ultimately to the conception of humanity itself, also becomes increasingly apparent. Competition and survival of the fittest are never wholly eliminated, but reappear on each new plane to work out the predominance of the higher, *i.e.*, more integrated and associated type, the phalanx being victorious till in turn it meets the legion. But this service no longer compels us to regard these agencies as the essential mechanism of progress, to the practical exclusion of the associative factor upon which the victory depends, as economist and biologist have too long misled each other into doing. For we see that it is possible to interpret the ideals of ethical progress, through love and sociality, co-operation and sacrifice, not as mere utopias contradicted by experience, but as the highest expressions of the central evolutionary process of the natural world. The ideal of evolution is indeed an Eden; and although competition can never be wholly eliminated, and progress must approach without ever completely reaching its ideal, it is much for our pure natural history to recognize that 'creation's final law' is not struggle but love." [40]

It is these views which it will fall to the lot of twentieth century investigators to substantiate.

Let us hear what some distinguished biologists of our time have to say upon these matters.

SIMPSON'S CONCEPTION OF EVOLUTION

In what is undoubtedly one of the best general works on evolution published in our time, *The Meaning of Evolution,* Professor George Gaylord Simpson, writes: "Natural selection as it was understood in Darwinian days emphasized 'the struggle for existence' and 'the survival of the fittest.' These concepts had ethical, ideological, and political repercussions which were and continue to be, in some cases, unfortunate, to say the least. Even modern students of evolution have not always fully corrected the misconceptions arising from these slogans. It should now be clear that the process does not depend on 'existence' or 'survival,' certainly not as this applies to individuals and not even in any intensive or explanatory way as it applies to populations or species. It depends on differential reproduction, which is a different matter

---

[40] Patrick Geddes and J. Arthur Thomson, *The Evolution of Sex,* Scott, London, 1889, pp. 311-312.

altogether. It does not favor the 'fittest,' flatly and just so, unless you care to circle around and define 'fittest' as those that do have most offspring. It does favor those who have more offspring. This usually means those best adapted to the conditions in which they find themselves or those best able to meet opportunity or necessity for adaptation to other existing conditions, which may or may not mean that they are 'fittest,' according to understanding of that word. Moreover the correlation between those having more offspring, and therefore really favored by natural selection, and those best adapted or best adapting to change is neither perfect nor invariable; it is only approximate and usual.

"It is, however, the word 'struggle' that has led to most serious misunderstanding of the process of natural selection, along with a host of related phrases and ideas, 'nature red in fang and claw,' 'class struggle' as a natural and desirable element in societal evolution, and all the rest. 'Struggle' inevitably carries the connotation of direct and conscious combat. Such combat does occur in nature, to be sure, and it may have some connection with differential reproduction. A puma and a deer may struggle, one to kill and the other to avoid being killed. If the puma wins, it eats and presumably may thereby be helped to produce offspring, while the deer dies and will never reproduce again. Two stags may struggle in rivalry for does and the successful combatant may then reproduce while the loser does not. Even such actual struggles may have only slight effects on reproduction, although they will, on an average, tend to ex-

ercise some selective influence. The deer most likely
to be killed by the puma is too old to reproduce; if the
puma does not get the deer, it will eat something else;
the losing stag finds other females, or a third enjoys
the does while the combat rages between these two.

"To generalize from such incidents that natural
selection is over-all and even in a figurative sense
the outcome of struggle is quite unjustified under
the modern understanding of the process. Struggle
is sometimes involved, but it usually is not, and when
it is, it may even work against rather than toward
natural selection. Advantage in differential reproduction is usually a peaceful process in which the
concept of struggle is really irrelevant. It more often
involves such things as better integration into the
ecological situation, maintenance of a balance of nature, more efficient utilization of available food, better care of the young, elimination of intragroup dis-
cords (struggles) that might hamper reproduction,
exploitation of environmental possibilities that are
not the objects of competition[41] or are less effectively
exploited by others." [42]

---

[41] [Simpson's note.] "The word 'competition,' used in discussion here and previously, may also carry anthropomorphic undertones and then be subject to some of these same
objections. It may, however, and in this connection it must,
be understood without necessary implication of active
competitive behavior. Competition in evolution often or
usually is entirely passive; it could conceivably occur without the competing forms ever coming into sight or contact."

[42] George G. Simpson, *The Meaning of Evolution,* Yale
University Press, New Haven, Conn., 1949, pp. 221-222.
Now available in revised and condensed form as a Men-
tor Book, New York, 1951.

Simpson goes on to point out that the group of its own kind among which an animal lives is also a part of its environment, but a special part. "There is an intraspecific selection, based on integration and association within the group, as well as extraspecific selection, based on adaptive relationship to the environment as a whole. . . . Intragroup selection may involve actual struggle, as in the case of the stags fighting for a doe. It may then be deleterious as regards extragroup adaptation and involve selection opposed to extragroup selection. If such is the case, the result, as Haldane[43] has emphasized, may be deleterious for the species as a whole, although even here we may remark that intra- and extragroup struggle commonly produce selection in the same direction. It is to be added that in intragroup selection, also, struggle is not necessarily or even usually of the essence. Precisely the opposite, selection in favor of harmonious or co-operative group association, is certainly common.

---

[43] J. B. S. Haldane, *The Causes of Evolution*, Longmans, New York, 1935, pp. 125-126. "It is in the struggle between adults of the same species that the biological effects of competition are probably most marked. It seems likely that they render the species as a whole less successful in coping with its environment. No doubt weaklings are weeded out, but so would they be in competition with the environment. And the special adaptations favoured by interspecific competition divert a certain amount of energy from other functions, just as armaments, subsidies, and tariffs, the organs of international competition, absorb a proportion of the national wealth which many believe might be better employed."

"It was a crude concept of natural selection to think of it simply as something imposed on the species from the outside. It is not, as in the metaphor often used with reference to Darwinian selection, a sieve through which organisms are sifted, some variations passing (surviving) and some being held back (dying). It is rather a process intricately woven into the whole life of the group, equally present in the life and death of the individuals, in the associative relationships of the population, and in their extra-specific adaptations." [44]

DOBZHANSKY DEFINES "ADAPTIVE VALUE"

Professor Theodosius Dobzhansky, one of the leading geneticists of our time, writes as follows: "Evolution is utilitarian in the sense that organisms change in the process of becoming adapted to their environments. The adaptation is brought about by natural selection which, in turn, is the outcome of differential perpetuation of different genotypes. Differential perpetuation is often styled 'competition' and 'struggle for life.' Both expressions are metaphors, and have often been misconstrued. Imagine two species of bacteria or two genetic types of the same species of bacteria which multiply in the same test tube with nutrient broth. They are 'competing' with each other in the sense that the more food one of them consumes, the less is left for the other. But the bacteria do not devour each other. When two species or varieties of grass occur in the same meadow they 'strug-

---

[44] Simpson, op. cit., p. 223.

gle' with each other, in the sense that there is only a limited amount of space available for their growth. But this struggle does not involve anything like fighting in the human sense. *Competition* and *struggle* are emotionally loaded words, which are best avoided in discussions of causes of evolution.

"No less misleading is the expression 'survival of the fittest,' which Herbert Spencer unfortunately coined to describe the operation of natural selection, and which became associated with something like the image of the Nietzschean superman. Now, *fitness,* in the evolutionary sense, or *adaptive value,* as it is better called, does not necessarily connote even a superior ability of an individual to survive, and a lack of fitness in this sense is not synonymous with weakness or frailty. A superior adaptive value of one genotype over another simply means that the carriers of the former leave, on the average, more surviving progeny than do the carriers of another genotype in the same environment. The superiority may result from the fact that individuals of one genetic type are stronger and more resistant to environmental hazards, and live longer than individuals of other genetic types. Or one type may be more sexually active or more fecund than another. Individual vigor and fecundity are not necessarily correlated, and a superior fecundity may compensate or even overcompensate for deficient vigor."[45]

---

[45] Th. Dobzhansky, "Heredity, Environment, and Evolution," *Science,* vol. III, 1950, pp. 161-166.

"FITNESS": A RELATIVE FUNCTION

The phrase "survival of the fittest", introduced by Herbert Spencer and adopted by Darwin, was a most unfortunate one, as Thomas Henry Huxley realized. In a letter written in 1890 he commented upon the fact that "The unlucky substitution of 'survival of the fittest' for 'natural selection' has done much harm in consequence of the ambiguity of 'fittest'—which many take to mean 'best' or 'highest'—whereas natural selection may work towards degradation *vide epizoa*." [46] By *epizoa* Huxley meant the category of animals that live as parasites upon the exterior of the bodies of other animals. Such parasitic animals are frequently characterized by an extreme loss of structural complexity. In the earlier stages of their individual development they may be quite complexly organized creatures; in their adult stages their structural organization may be reduced to the simplest form. *Sacculina,* the crustacean that belongs to the group of barnacles and attaches itself to the body of crabs, is a good example of such a parasite.

T. H. Huxley's criticism of "survival of the fittest" was very much to the point. What many biologists and most Social Darwinists overlooked was the fact that "fitness" is related to the current environment of a group and is not either necessarily or usually a capacity or function that enables the group or the

---

[46] *Life and Letters of Thomas Henry Huxley* (edited by Leonard Huxley), Appleton, New York, 1901, vol. 2, p. 284.

organism to adapt itself to *all* environments. Failure
to recognize this fact led to the conversion of "fittest"
into "best" under all conditions. The fact is that "fitness" is a relative function, a function of the organism in relation to a particular environment.

INTERDEPENDENCE OF ORGANISMS

Dr. Marston Bates, Professor of Zoology at the University of Michigan, in his recent book *The Nature
of Natural History*, writes, "The point I am trying to make here is the interdependence of kinds
of organisms in the world today. We have got into
the habit of looking at the organic world as a mass
of struggling, competing organisms, each trying to
best the other for its place in the sun. But this competition, this 'struggle,' is a superficial thing, super-
imposed on an essential mutual dependence. The
basic theme in nature is cooperation rather than competition—a cooperation that has become so all-per-
vasive, so completely integrated, that it is difficult to
untwine and follow out the separate strands." [47]

"I think," writes Dr. Bates, "there has been an increasing tendency in biological writing to stress the
cooperative rather than the competitive aspects of
relations among various kinds of organisms. But this
tendency has not been adequately reflected in the
thinking of the social philosophers, who have tended
to confine their biological explorations to the post-
Darwinian, or Thomas Huxley period. I don't think

[47] Marston Bates, *The Nature of Natural History*, Scribner, New York, 1950, p. 108.

the social philosophers are entirely to be blamed for this. The biologists have failed to make their growing knowledge, their accumulating facts and concepts, easily available to the philosophers." [48]

HARMONIOUS ADJUSTMENT AND ALTRUISTIC TENDENCIES

Professor Warder C. Allee of the University of Chicago and his four collaborators in their great work *Principles of Animal Ecology,* published in 1949, conclude this magnificent volume with the following words: "We may thus summarize the section on Ecology and Evolution—and indeed the book as a whole—by repeating a principle discussed by Leake (1945): The probability of survival of individual living things, or of populations, increases with the degree with which they harmoniously adjust themselves to each other and their environment. [49] This principle is basic to the concept of the balance of nature, orders the subject matter of ecology and evolution, underlies organismic and developmental biology, and is the foundation for all sociology." [50]

In an earlier publication Professor Allee has written: "After much consideration, it is my mature con-

---

[48] *Ibid.*, p. 123. See Bates' Chapter IX, "Partnership and Cooperation," pp. 122-136.

[49] For the development and critical discussion of this principle see Chauncey D. Leake and Patrick Romanell, *Can We Agree?* University of Texas Press, Austin, 1950.

[50] Warder C. Allee, Alfred E. Emerson, Orlando Park, Thomas Park, and Karl P. Schmidt, *Principles of Animal Ecology,* Saunders, Philadelphia, 1949, p. 729.

clusion, contrary to Herbert Spencer, that the co-operative forces are biologically the more important and vital. The balance between the co-operative and altruistic tendencies and those which are disoperative and egoistic is relatively close. Under many conditions the co-operative forces lose. In the long run, however, the group centered, more altruistic drives are slightly stronger. If co-operation had not been the stronger force, the more complicated animals, whether arthropods[51] or vertebrates, could not have evolved from simpler ones, and there would have been no men to worry each other with their distressing and biologically foolish wars. While I know of no laboratory experiments that make a direct test of this problem, I have come to this conclusion by studying the implications of many experiments which bear on both sides of the problem and from considering the trends of organic evolution in nature. Despite many known appearances to the contrary, human altruistic drives are as firmly based on animal ancestry as is man himself. Our tendencies toward goodness, such as they are, are as innate as our tendencies toward intelligence; we could do well with more of both." [52]

---

[51] The phylum of arthropods (*Arthro*, Greek, meaning joint, *podos*, a foot, literally animals with jointed feet or legs) includes more than half the known species of animals and embraces such creatures as the insects, lobsters, crayfish, centipedes, millipedes, and spiders.

[52] Warder C. Allee, "Where Angels Fear to Tread: a Contribution from General Sociology to Human Ethics," *Science*, vol. 97, 1943, p. 521.

And, again, Professor Allee points out that "Today, as in Darwin's time, the average biologist apparently still thinks of a natural selection which acts primarily on egoistic principles, and intelligent fellow thinkers in other disciplines, together with the much-cited man-in-the-street, cannot be blamed for taking the same point of view." [53]

In 1920 the late Professor William Patten, Professor of Biology at Dartmouth College, one of the most distinguished biologists of his day, published a book entitled *The Grand Strategy of Evolution*, and subtitled *The Social Philosophy of a Biologist*.[54] This book represents one of the most valuable contributions to theoretical biology of the twentieth century. It was very favorably noticed when it appeared, but is today little known. Possibly it has had more influence upon the minds of biologists than would appear from the paucity of references to it. Patten's book will always remain one of the most inspiring correctives to the Darwinian conception of evolution by natural selection through a misconceived competition. The book represents the personal testament of one who has thought long and deeply over the processes of nature. It constitutes the distillate of his thought and experience as a biologist. He concludes that "The universal end, or purpose in life, and in nature, is to construct, to create, or grow. The ways and means of accomplishing that end are mutual service, or co-operative action, and rightness.

---

[53] *Ibid.*, p. 520.

[54] William Patten, *The Grand Strategy of Evolution*, Badger, Boston, 1920.

"The initial constituents of any growing thing, by acting the more cooperatively, the better preserve themselves. These in turn cooperate to create and preserve other constructive agents. But these creative and saving acts cannot take place unless the right things are conveyed to the right place at the right time. When these services are performed, the creative and saving acts take place spontaneously, or automatically. . . .

"In all these cases, the reciprocal egoism and altruism, the creating and saving of self in order to give self to some other creative act, is the essential process in evolution. It is the perpetual source of nature's increasing supplies and of her insatiable demands. It is the basis of man's system of ethics and morality, for in human affairs we call these same constructive processes, righteousness, cooperation, altruism, service, benevolence, and self-sacrifice." [55]

NATURALISTS PAINT A ONE-SIDED PICTURE

"The picture of nature painted by the field naturalists was a warring, hostile nature, 'red in tooth and claw with ravin.' Its merciless 'struggle for existence,' its wanton destruction and tragic incident deeply moved both scientist and layman and greatly influenced the conduct and the interpretation of human life.

"But the attention of the naturalists was chiefly focused on the fifth act of life's drama, not on the body of the play; on the inevitable collapse of imperfect life systems, not on the upbuilding of cell on cell,

[55] *Ibid.*, pp. xii-xiii.

and life on life, during long preliminary constructive periods.

"They gave us the picture of life's tragic side, and,
like all one-sided representations, it was but a carica-
ture, true, indeed, to the life they portrayed, but
misleading in the omission of the larger truth. They
showed us the shameless selfishness, the fruitless toil
and suffering, the wanton wastefulness of life and
the endless competition in strength and skill, in shift-
ing alliance with hypocrisy and deceit; with blind
chance in the background awarding death to the
vanquished and to the victor life's bitter spoils. With
master strokes, and with the convincing accuracy
of the trained observer, they painted the 'disastrous
chances' of a tumultuous life, 'the moving accidents
by flood and field,' the spectacular catastrophes of
failure.

"But they did not portray the benevolent processes
of construction, the peaceful cooperation, the careful conservation, and the successful sacrifice of self
to higher service. Some writers have indeed recognized the element of benevolence in the cooperation
that forms such a conspicuous feature of many social organizations; but usually it has been regarded
as something peculiar to a few highly organized animals and to man, not as something inherent to all
stages of organic and inorganic nature." [56]

"Natural selection and the survival of the fittest
are perhaps the broadest terms used in the biological
sciences, but the processes so designated have no
creative value. The terms merely imply that a defi-

[56] *Ibid.*, pp. 25-26.

nite sequence of products ensues, or affirm the self-evident fact that something already created is selected for survival, or that it endures. They do not suggest how it was created, why it survives, or wherein its fitness lies.

"I shall try to show that there is but one answer to all these questions; that there is but one creative process common to all phases of evolution, inorganic, organic, mental, and social. That process is best described by the term cooperation, or mutual service." [57]

I wish I could go on quoting Patten at length, but to do so would be virtually to quote his whole book. I hope the excerpts I have already given from it will encourage the reader to seek out the book itself.

Concluding a great work on evolution, the late John Muirhead Macfarlane, Professor of Botany at the University of Pennsylvania, wrote, "Competition then, as a fundamental zoological law, is not nearly so 'successful,' we believe, as that of cooperation or social union." [58]

ALTRUISM AS PASSION

Another quotation, this time from the pen of one of the greatest of physiologists, Sir Charles Sherring-

---

[57] *Ibid.,* pp. 32-33.

[58] John M. Macfarlane, *The Causes and Course of Organic Evolution,* Macmillan, New York, 1918, p. 755. See Macfarlane's two chapters, (27) "The Competitive System Amongst the Lower Animals and Man," and (28) "The Cooperative or Social System Amongst Lower Animals and Man."

ton (1861-1952), from one of the greatest books of our time. "The individual?" he asks. "What are the most successful individuals which life has to show? The multi-cellular. And what has gone to their making? The multi-cellular organism is in itself a variant from the perennial antagonism of cell and cell. Instead of that eternal antagonism it is a making use of relatedness to bind cell to cell for co-operation. The multi-cellular organism stood for a change, in so far, from conflict between cell and cell to harmony between cell and cell. Its coming was, we know now, pregnant with an immense advance for the whole future of life upon the globe. It was potential of the present success of living forms upon the planet. Implicit was for one thing the emergence of recognizable mind. It was among the many-celled organisms that recognizable mind first appeared. It is surely more than mere analogy to liken to those small beginnings of multi-cellular life of millions of years ago the slender beginnings of altruism today. Evolution has constantly dealt with the relation between physical and mental as more than mere analogy. The bond of cohesion now arising instead of being as then one of material contact between related cell-lives is in its nature mental. It is a projection of the self by sympathy with other life into organismal situations besides its immediate own. It is altruism as passion." [59]

---

[59] Charles Sherrington, *Man on His Nature,* Cambridge University Press, New York, 1st ed., 1941, pp. 387-388; 2d ed., 1951.

THE PROPERTIES OF GOOD AND EVIL

Professor Ralph Lillie, Professor Emeritus of General Physiology at the University of Chicago, has pointed out that "a property of the *good* (in the universal or Platonic sense) is that conscious tends to be directed toward its continuance, since it is the object of desire; while *evil,* the immediately or ultimately painful, is a feature of reality which conscious effort tends to remove or overcome. The former has thus within itself a property or character which favors its continuance and increase; the latter is inherently unstable.

"Scientific analysis shows that stability in all highly diversified or composite systems requires harmonious relations—relations of mutual support or equilibrium —between the different components and activities . . . what should be better known and more widely acted upon is that integration *between* different types of individuals, as seen in the mutually helpful relations of the various units in many human and animal communities—or even between different species of animals and plants—is as much a factor in biological survival and evolution as is conflict. The avoidance of useless conflict, and the subordination of individual interests to the interest of the whole reality which includes the individuals, would thus seem to be rational aims for all conscious beings; and these aims have the further sanction of religion when the

whole is conceived in its character as ultimate value
or deity." [60]

THE URGE TOWARD UNDERSTANDING AND LOVE

Finally, let us listen to the words of one of the
world's outstanding geneticists, Professor Herman J.
Muller, who, in his presidential address before the
Eighth International Congress of Genetics, held at
Stockholm, July 7, 1948, said, "What then have been
the characteristics which above all have made man
great, in our deepest inner estimations, and to the
furtherance of which we should be the most willing
to devote ourselves? Certainly, for one thing, his intelligence. We would find no point in working to
bring about a world of witless giant machines alone,
nor of lumbering brutes that had no psychic future,
whereas even the fanciful idea of intelligences that
are disembodied has for ages exercised some attrac-
tion for us. Man's intelligence however has won its
victories chiefly because it was a cooperative one, and
it would not have been such had there been no strong
fellow-feeling. It is the attachment of mother for
child, of man for woman, and of man for man, that
have bound us into the little groups that won, and
that are finally binding us into the universal group,
and it is this basic feeling of love that will continue
to make the living world go round, if go it does. It is
this for which we most readily put forth sacrifices.

---

[60] Ralph S. Lillie, *General Biology and Philosophy of
Organism*, University of Chicago Press, Chicago, 1945,
pp. 208-209.

"There may well be still further developments in biological evolution, of quite different kinds, but, if we knew of them now, they could have a salutary significance for us only in so far as they ministered, directly or indirectly, to these two central trends of our natures: the urge towards understanding and towards love. I cannot see farther than this: that the end of all human striving must be for the strengthening and increase of these two aspects of our being. They need for us no external sanction: we are made so that they themselves constitute our justification for all things and thoughts and acts which serve them, for by serving them we serve life and by serving life we serve them. Joy of life, appreciation of the external world, are means towards their magnification. Power, in the form either of control over the environment or of control over our inner natures, this is for us a tool for attaining them, whereby we may work towards them in both their quantitative and qualitative aspects. It is for them, ultimately, that we must attempt to conquer all other things, including other living things. As between the two of them, we can make no comparison that would set either one above the other, for they are both the indispensable legs of the one body, intelligent cooperation, which raises man so far above all other life forms that we know of. Observe moreover that they themselves form no static end, being also the means of striving towards their own ever greater and higher development. Like all aspects of life, they are cyclic and self-multiplying: the end becomes a means, and the means again leads to this end, in an unlimited rotation that becomes greater and more manifold as it proceeds.

"At present, we have no way of telling how far this can go, but if the progression cannot be infinite it can yet be so great, in relation to the individual man, as practically to suffice him. But only when a man is taught this ideal from his early childhood years can it capture him in full measure, and lead him to the greatest richness, and the greatest usefulness, of his own life. Similarly too, a society as a whole can advance most rapidly only if founded in acknowledgment of this ideal as its basis. Religions of the past have derived much of their success by catching at parts of these truths, and much of their failures by grievously missing other parts." [61]

DARWINIAN "NATURAL SELECTION": A FRUITFUL ERROR

In all that has thus far been said it should be clear that the greatness of Darwin's achievement is being neither minimized nor denied. What is denied is its soundness, since it fails to tell the whole story. An incomplete or even a wrong theory can be the starting-point of a sound one. And, indeed, the development of science could well be written in terms of a history of fruitful errors. Darwin's conception of natural selection is perhaps the most outstanding example in the whole history of science of such a fruitful

[61] Herman J. Muller, "Genetics in the Scheme of Things," *Hereditas*, Supplementary Volume, 1949, pp. 113-114.

error. It is a conception which is only partially erroneous, firstly because it overemphasized the role of competition, and, secondly, because it failed to recognize the importance of co-operation in evolution.

Natural selection is a sound enough principle, taken in the round. It has been operative in determining the fate of *populations* of organisms since the beginning of life. There is evidence, however, that it has been rather more operative in terms of co-operation, than it has been in terms of what is generally understood by competition.

Today, contrary to the "nature, red in tooth and claw" school of natural selectionists, the evidence increasingly indicates that natural selection does not act principally to favor variations which through a ruthless kind of competition better adjust the organism to its environment. Adjustment is, of course, necessary, but the important point is that natural selection favors the co-operative, as opposed to the disoperative, struggle for life. To "struggle for life" is to be co-operative, for life is of its nature social and all activities calculated to maintain it in the individual and in the species are co-operative. As Delage and Goldsmith put it, "Natural selection always asserts itself and is a mighty factor, but how does its action make itself felt? Through the survival of those who know best how to make use of their aptitude for social life, which, in the universal struggle, becomes one of the most efficient weapons." [62]

---

[62] Yves Delage and Marie Goldsmith, *The Theories of Evolution*, Huebsch, New York, 1912, pp. 350-351.

This does not mean that under natural conditions the process of life does not involve some relationships which may be largely or exclusively ruthlessly competitive. Natural selection frequently involves some kind of "competition," but this competition need be no more ruthless than the competing of boys for a prize or persons sitting in a room unconsciously competing with each other or as a group with other groups for the air in the room which it is necessary for them to breathe if they are to survive. Furthermore, the type that ultimately contributes its gene complex to succeeding generations may have been more successful than its "competitors" because the requirements of the situation, over the course of biological time, placed a premium upon co-operation. From this point of view, then, co-operation would be said to have greater adaptive value in "the struggle for life."

Professor Paul R. Burkholder concludes a notable study of this subject with the following words, "The most important basis for selection is the ability of associated components to work together harmoniously in the organism and among organisms. All new genetic factors, whether they arise from within by mutation or are incorporated from without by various means, are accepted or rejected according to their co-operation with associated components in the whole aggregation." [63]

We begin to understand, then, that evolution itself is a process which favors co-operating rather than disoperating groups, and that "fitness" is a function

[63] P. R. Burkholder, "Co-operation and Conflict Among Primitive Organisms." Ms.

of the group as a whole rather than of separate individuals. The fitness of the individual is largely derived from his membership in the group.

In so far as man is concerned, if competition, in its aggressive combative sense, ever had any adaptive value among men, which is greatly to be doubted, it is quite clear that it has no adaptive value whatever in the modern world. Today the adaptive value of human aggression is so low as to threaten the existence of all the members of groups exhibiting it. Perhaps never before in the history of man has there been so high a premium upon the adaptive value of co-operative behavior, of peace on earth and goodwill toward all men.

Natural selection through "competition" may secure the immediate survival of certain types of "competitors," but the survivors would not long survive if they did not co-operate. Darwin regarded natural selection as equivalent to preservation. That is evident from the title of his book: *On the Origin of Species by Means of Natural Selection, Or the Preservation of Favoured Races in the Struggle for Life*. It is also clear from a letter to H. G. Bronn in which he writes, "Man has altered, and thus improved the English race-horse by *selecting* successive fleeter individuals; and I believe, owing to the struggle for existence, that similar *slight* variations in a wild horse, *if advantageous to it*, would be *selected* or *preserved* by nature; hence, Natural Selection." [64]

---

[64] *The Life and Letters of Charles Darwin* (edited by Francis Darwin), John Murray, London, 1888, vol. 2, p. 279, letter dated February 14 [1860]. See also *More*

Again in August, 1860, Darwin writes, "If I had to rewrite my book, I would use 'natural preservation' or 'naturally preserved.' " [65]

Although it may not have been obvious that organisms are *not* selected for survival by means of natural selection through "competition," it should have been clear that they could not be *preserved* by any other means than co-operation.

SOCIAL INSTINCTS OF ANIMALS

Now, the interesting thing is that although Darwin prepared six editions of *The Origin of Species,* the sixth and last edition appearing in January, 1872, a year after the publication of *The Descent of Man* (which appeared in 1871), Darwin nowhere in the *Origin* exhibits any evidence of having recognized even so much as the existence of co-operation as a factor of evolution. Neither is the notion to be found anywhere in his letters. Yet in *The Descent of Man* the idea of co-operation as a factor in the evolution of man, at least, is granted more than a merely equivocal existence, so far, at any rate, as man is concerned.

So firmly were Darwin's preconceptions ingrained in his mind that when he himself, with his beautiful clarity, practically describes the origin of co-operation he fails to understand the full significance of what

---

Letters of Charles Darwin (edited by Francis Darwin and A. C. Seward), John Murray, London, 1903, vol. 1, p. 126.

[65] *Ibid.,* vol. 1, p. 161.

he says. Thus, he writes, "It has often been assumed that animals were in the first place rendered social, and that they feel as a consequence uncomfortable when separated from each other, and comfortable whilst together; but it is a more probable view that these sensations were first developed, in order that those animals which would profit by living in society, should be induced to live together, in the same manner as the sense of hunger and the pleasure of eating were, no doubt, first acquired in order to induce animals to eat. The feeling of pleasure from society is probably an extension of the parental or filial affections, since the social instinct seems to be developed by the young remaining for a long time with their parents; and this extension may be attributed in part to habit, but chiefly to natural selection. With those animals which were benefited by living in close association, the individuals which took the greatest pleasure in society would best escape various dangers, whilst those that cared least for their comrades, and lived solitary, would perish in greater numbers. With respect to the origin of the parental and filial affections, which apparently lie at the base of the social instincts, we know not the steps by which they have been gained; but we may infer that it has been to a large extent through natural selection." [66]

Had Darwin not been so much under the influence of the "struggle for existence" view of life, he might have glimpsed the truth that no animal is, in fact,

---

[66] *The Descent of Man,* Chap. IV, p. 161.

solitary.[67] He might have seen the fact that some form of social life is characteristic of all living things, and he might have been led on to discover that preservation depends not upon competition, but upon co-operation. And hence his whole theory of natural selection would have been different.[68] But the con-

---

[67] "There is something fundamentally social in living things, and closer scrutiny shows that this must be a characteristic of all life, since every organism is, at least temporarily, associated with other organisms, even if only with members of the opposite sex and with its parents. . . . This statement holds good even of such supposedly unsocial creatures as lions, eagles, sharks, tiger-beetles, and spiders. There are, in fact, no truly solitary organisms. We may say, therefore, that the social is a correlate as well as an emergent of all life in the sense in which Morgan speaks of the mind as being both a correlate and an emergent of life. . . . Indeed, the correlations of the social—using the term in its most general sense—even extend down through the non-living to the very atom with its organization of component electrons." William M. Wheeler, "Emergent Evolution and the Development of Societies," in *Essays in Philosophical Biology,* Harvard University Press, Cambridge, Mass., 1939, pp. 158-159. "A solitary individual wholly independent of others is largely a fiction. In reality, most or even all living beings exist in more or less integrated communities, and the ability to maintain these associations entails some cooperation, or at least 'protocooperation.'" Th. Dobzhansky, *Genetics and the Origin of Species,* 3d ed., Columbia University Press, New York, 1951, pp. 78-79.

[68] For excellent general criticisms of the concept of natural selection from the standpoint of the modern biologist, see Lancelot Hogben, "Natural Selection and Re-

cept of "if" implies conditions, and the necessary conditions for the development of a less brutal theory do not seem to have been present to his mind.

THE CONCEPT OF MAN AS A BRUTE

One may say of Darwin, as Lancelot Hogben has said of Huxley and Spencer, that to him, Darwin, "the important fact was that Man is a brute. It was necessary for [him] to emphasize man's genetic similarity to other animals in opposition to the traditional view which placed man in a special category apart from other natural objects." [69] Indeed, in *The Descent of Man* Darwin seems almost deliberately to go out of his way to make the point that man's pedigree consists of something less than a long line of cherubim. Thus, he writes, ". . . we have given man a pedigree of prodigious length, but not, it may be said, of

---

search," in *The Nature of Living Matter*, Knopf, New York, 1931, pp. 170-190, and Samuel J. Holmes, "What Is Natural Selection?" *Scientific Monthly*, vol. 67, 1948, pp. 324-330. See also Herman J. Muller, "The Darwinian and Modern Conceptions of Natural Selection," *Proceedings of the American Philosophical Society*, vol. 93, 1949, pp. 460-471; Th. Dobzhansky, *Genetics and the Origin of Species*, 3d ed., Columbia University Press, 1951; George G. Simpson, *The Meaning of Evolution*, Yale University Press, New Haven, 1949, Mentor Books, New York, 1951; I. I. Schmalhausen, *Factors of Evolution: The Theory of Stabilizing Selection*, Blakiston, Philadelphia, 1949.

[69] Lancelot Hogben, *The Nature of Living Matter*, Knopf, New York, 1931, p. 194.

noble quality." [70] Such a statement may throw light upon the cast of Darwin's thought, but to some it may seem both gratuitous and otiose. Theological opposition to evolutionary thought had produced a certain truculent reaction in the pioneers of evolution, and I fancy we may detect a faint reflection of it in Darwin's remark concerning man's pedigree.

SOCIAL INTERACTION: ITS ORIGIN

What Darwin in the bleak age of struggle and imperialism (imperialism being but another name for the doctrine of the survival of the fittest peoples) failed to perceive was the dependence of cell upon cell, of organ upon organ, of organism upon organism. [71]

---

[70] *The Descent of Man,* Chap. VI, p. 255.

[71] Lest it be thought that in Darwin's case this was so because of his unfamiliarity with cellular physiology or embryogenetics, it may be mentioned that this failure also affected the great founder of experimental embryology, Wilhelm Roux (1850-1924). Roux, under the influence of the Darwinian conception of the struggle for existence, published a book entitled *Der Kampf der Teile im Organismus* [*The Struggle of the Parts Within the Organism*] (1881) in which he attempted to show that a struggle is constantly going on within every organism between the organic molecules, the cells, and the organs of which it is constituted. Finding that the cells which best perform their functions are the best nourished and also best propagate their kind, he concluded that natural selection through competition also

The evidence shows that all living things exhibit some form of social behavior, that is to say, a process of interaction with other organisms, usually of their own kind, which has the effect of satisfying their mutual needs, and is thus productive of survival benefits for the organism and the group. As Holmes has recently put it, "Social groups, whether of cells or loose aggregates of human nomads, have much the same biological significance. They are mutual benefit associations and are dependent upon a measure of ac-

---

plays an important role in the economy of the organism. The difference in point of view, and hence in interpretation, may be illustrated by the following brief example. The lungs expand with atmospheric air at birth and continue thereafter to enlarge and press against the heart. One may speak of this as resulting in a "competition for space," a "struggle for supremacy" between these organs. Alternatively one may describe the changes as leading to an interdependent adjustment or a co-operative adjustment between the organs involved. Much to the point here is the recent comment of Professor Herbert W. Rand who, in a recent review of W. K. Gregory's great work *Evolution Emerging*, writes as follows: "The author's discussion of the evolutionary status of *Amphioxus* (p. 88) invites comment. He depicts *Amphioxus* as an unfortunately frustrated little creature which might have had a bigger and better brain had not the front end of the notochord been in the way. This calls to mind Wilhelm Roux and his *Kampf der Teile* (1881). *But a half century of experimental embryology teaches us that the conspicuous feature of embryonic development is not competition but cooperation or coordination.*" (My italics.) [72]

[72] "The Cosmic Cinema," *Science*, vol. 113, 1951, p. 437.

tivity which is altruistic in effect, whether or not it is consciously so in intent." [73]

This tendency to social interaction may have some deeper source, but it appears to me to have its origin in the fact that all organisms come into being from other organisms. Whether they are reproduced asexually or sexually, all organisms in the process of coming into being are for a time dependent upon and interdependent with the organism from which they are being generated. Whatever happens to the one affects the other. The needs of the one are satisfied by the other—thus, the origin of interdependency. Furthermore, the cell which is coming into being is for an appreciable period dependent upon the maternal cell. It is suggested that it is in this process of dependency and interdependency, the reproductive process, that the meaning of the origin and nature of social life, of co-operation, is to be sought. It is in the very nature of living tissues to be interactive, to strive to be social. Again, as Holmes has stated, ". . . antecedent to the activities that we ordinarily designate as behavior, in the sense in which this term is employed in comparative psychology, there are the physiological processes of producing, storing, and discharging germ cells, and in viviparous animals, carrying and nourishing the young. These processes, though mainly unconscious, are nevertheless altruistic in that they conduce to maintaining life in another individual. Altruism has emerged, starting with the fission of the simplest organism; then it is ex-

[73] S. J. Holmes, *Life and Morals*, Macmillan, New York, 1948, p. 4.

pressed in more complex ways in the reproductive activities of higher forms, and later manifested in overt behavior with its accompanying sensory and emotional experiences. Naturally its first manifestation occurs in the family. As the family expands into the larger social group, unselfish aid is given to other individuals of the community." [74]

MAN'S SUPERIORITY AND SOCIAL INSTINCTS

Such ideas do not seem to have occurred to Darwin, though there are occasions in *The Descent of Man* in which there is explicit recognition of the value of co-operation, of altruism, in man. Thus, in his first reference to the subject Darwin writes, "Man in the rudest state in which he now exists is the most dominant animal that has ever appeared on this earth. He has spread more widely than any other highly organised form: and all others have yielded before him. He manifestly owes this immense superiority to his intellectual faculties, to his social habits, which lead him to aid and defend his fellows, and to his corporeal structure. The supreme importance of these characters has been proved by the final arbitrament of the battle for life. Through his powers of intellect, articulate language has been evolved; and on this his wonderful advancement has mainly depended." [75]

This passage could be taken to refer to co-operation within the group rather than to co-operation extended to members of outside groups. This may be

---

[74] *Ibid.*, p. 113.
[75] *The Descent of Man,* Chap. II, p. 72.

its meaning. In which case we may again quote a passage from Darwin in which his meaning is un-equivocally clear. "Notwithstanding many sources of doubt, man can generally and readily distinguish be-tween the higher and lower moral rules. The higher are founded on the social instincts, and relate to the welfare of others. They are supported by the appro-bation of our fellow men and by reason. The lower rules, though some of them when implying self-sacrifice hardly deserve to be called lower, relate chiefly to self, and arise from public opinion, matured by experience and cultivation; for they are not prac-tised by rude tribes." [76]

"As man advances in civilisation, and small tribes are united into larger communities, the simplest rea-son would tell each individual that he ought to ex-tend his social instincts and sympathies to all the members of the same nation, though personally un-known to him. This point being once reached, there is only an artificial barrier to prevent his sympathies extending to the men of all nations and races. If, indeed, such men are separated from him by great differences in appearance or habits, experience un-fortunately shews us how long it is, before we look at them as our fellow-creatures." [77]

Now, when Darwin says that the higher moral rules "are founded on the social instincts"—which he defines as a form of behavior "that is followed inde-

---

[76] This sentence is quite incorrect, as Kropotkin, in *Mutual Aid*, was among the first to show, and as modern anthropology has abundantly demonstrated.

[77] *The Descent of Man*, Chap. IV, pp. 187-188.

pendently of reason"—and "relate to the welfare of others," it is clear that he believes in an innate kind of altruism, at least, in man. Furthermore, he understands that altruism has played a role in the evolution of man. For example, he writes further, "It must not be forgotten that although a high standard of morality gives but a slight or no advantage to each individual man and his children over the other men of the same tribe, yet that an increase in the number of well-endowed men and an advancement in the standard of morality will certainly give an immense advantage to one tribe over another. A tribe including many members who, from possessing in a high degree the spirit of patriotism, fidelity, obedience, courage, and sympathy, were always ready to aid one another, and to sacrifice themselves for the common good, would be victorious over most other tribes; and this would be natural selection. At all times throughout the world tribes have supplanted other tribes; and as morality is one important element in their success, the standard of morality and the number of well-endowed men will thus everywhere tend to rise and increase." [78]

MORALITY DEPENDENT UPON COMPETITION

In other words, Darwin is here thinking of co-operation within the group—intra-group co-operation —as a factor of natural selection, and he is suggesting that co-operation is of high selective value, since it enables co-operative tribes to be victorious over most other tribes. His assumption is that "most other

[78] *Ibid.,* Chap. V, pp. 203-204.

tribes" were overcome because they had not achieved so high a degree of co-operation. Hence, morality itself is naturally selected. This kind of reasoning, that is, in terms of physical victory, is necessary to Darwin because he is constrained to think within the pattern of a natural selection which acts through the agency of a combative competition, through the struggle for survival, the survival of the fittest. Hence, he makes morality dependent not upon co-operation, which is a sort of desirable mutation, but upon competition—the mechanism which will determine its survival value—upon the victorious conquest of "most other tribes" by the tribes possessing a higher degree of altruism.[79] The members of each tribe must be altruistic among themselves, but to survive and evolve, according to Darwin, they must struggle with others. Thus, Darwin writes, "When two tribes of primeval man, living in the same country, came into competition, if (other circumstances being equal) the one tribe included a great number of courageous, sympathetic and faithful members, who were always ready to warn each other of danger, to aid and defend each other, this tribe would succeed better and conquer the other. Let it be borne in mind how all-important in the never-ceasing wars of savages,[80] fidelity and cour-

---

[79] This is a view which Sir Arthur Keith has recently developed at length in a somewhat anachronistic work entitled *A New Theory of Human Evolution*, Philosophical Library, New York, 1948. See also Keith's *Evolution and Ethics*, Putnam, New York, 1946.

[80] This is a gross overstatement. There are many so-called "savage" peoples who, so far as we know, have

age must be. The advantage which disciplined soldiers have over undisciplined hordes follows chiefly from the confidence which each man feels in his comrades. Obedience, as Mr. Bagehot has well shewn,[81] is of the highest value, for any form of government is better than none. Selfish and contentious people will not cohere, and without coherence nothing can be effected. A tribe rich in the above qualities would spread and be victorious over other tribes: but in the course of time it would, judging from all past history, be in its turn overcome by some other tribe still more highly endowed. Thus the social and moral qualities would tend slowly to advance and be diffused throughout the world." [82]

---

never engaged in warfare, and who at the present time scarcely understand its meaning—the Australian aborigines and the Eskimo, for example. For the facts see P. Chalmers Mitchell, *Evolution and War*, Dutton, New York, 1915; Havelock Ellis, *Essays in War-Time*, Constable, London, 1916; Maurice R. Davie, *The Evolution of War*, Yale University Press, New Haven, Conn., 1929; Quincy Wright, *A Study of War*, 2 vols., University of Chicago Press, Chicago, 1942; Harry H. Turney-High, *Primitive War*, University of South Carolina Press, Columbia, 1949; L. L. Bernard, *War and Its Causes*, Holt, New York, 1944.

[81] Walter Bagehot, *Physics and Politics, Or Thoughts on the Application of the Principles of Natural Selection and Inheritance to Political Society*, Appleton, New York, 1872. Darwin's reference to this work is to its appearance in the form of a series of articles in the *Fortnightly Review*, "since separately published."

[82] *The Descent of Man*, Chap. V, pp. 199-200.

The passage is of particular interest in view of the statement which has been made by numerous writers that "Darwin regarded war as an insignificant or even non-existent part of natural selection." [83] It would be difficult to interpret this and other passages already cited, and still others which cannot be cited here, as having any other meaning than that Darwin believed in warfare as an agency of natural selection in the evolution of man.

ORIGIN OF ALTRUISM

Before continuing with this subject let us see what Darwin really believed the origin of altruism to be. He writes, "Although the circumstances, leading to an increase in the number of those thus endowed within the same tribe, are too complex to be clearly followed out, we can trace some of the probable steps. In the first place, as the reasoning powers and foresight of the members became improved, each man would soon learn that if he aided his fellow-men, he would commonly receive aid in return. From this low motive he might acquire the habit of aiding his fellows; and the habit of performing benevolent actions certainly strengthens the feeling of sympathy which gives the first impulse to benevolent actions.

---

[83] Havelock Ellis, "Evolution and War," in *Essays in War-Time*, Constable, London, 1916, p. 16. See also Sir Norman Angell's introduction to George Nasmyth's *Social Progress and the Darwinian Theory*, Putnam, New York, 1916, and Nasmyth's own euhemerization of the myth of Darwin's essential "peacefulness."

Habits, moreover, followed during many generations probably tend to be inherited." [84]

Here again, the emphasis is on the "low motive." We, of course, today know that habits followed for howsoever many generations are not inherited.

ACTION OF NATURAL SELECTION ON MAN

Darwin's conception of the action of natural selection upon man would appear to be clear from the following passage, "Natural selection follows from the struggle for existence; and this from a rapid rate of increase. It is impossible not to regret bitterly, but whether wisely is another question, the rate at which man tends to increase; for this leads in barbarous tribes to infanticide and many other evils, and in civilised nations to abject poverty, celibacy, and to the late marriages of the prudent.[85] But as man suffers from the same physical evils as the lower animals, he has no right to expect an immunity from the evils consequent on the struggle for existence. Had he not been subjected during primeval times to natural selection, assuredly he would never have attained to his

---

[84] *The Descent of Man,* Chap. V, p. 201.

[85] For a brilliant modern criticism of this viewpoint, see Lancelot Hogben, "The Growth of Human Populations," in *Genetic Principles in Medicine and Social Science,* Knopf, New York, 1931, pp. 173-199. See also H. J. Muller, "Our Load of Mutations," *American Journal of Human Genetics,* vol. 2, 1950, pp. 111-176 and especially Josué de Castro, *The Geography of Hunger,* Little, Brown, Boston, 1952.

present rank. Since we see in many parts of the world enormous areas of the most fertile land capable of supporting numerous happy homes, but peopled only by a few wandering savages, it might be argued that the struggle for existence had not been sufficiently severe to force man upwards to his highest standard. Judging from all that we know of man and the lower animals, there has always been sufficient variability in their intellectual and moral faculties, for a steady advance through natural selection. No doubt such advance demands many favourable concurrent circumstances; but it may well be doubted whether the most favourable would have sufficed, had not the rate of increase been rapid, and the consequent struggle for existence extremely severe." [86]

Considering Darwin as a whole the reader, surely, cannot interpret such a passage as meaning anything other than that natural selection, the struggle for existence, in man largely involves conflict with other groups of men, those endowed with the "fittest" qualities emerging the victors.

THE EFFECTS OF CULTURAL EXPERIENCE

Darwin can hardly be criticized for not having understood why it is that "we see in many parts of the world enormous areas of the most fertile land capable of supporting numerous happy homes, but peopled only by a few wandering savages," for he wrote in an age before the development of modern cultural anthropology. What he attributed to the insufficiently

---

[86] *The Descent of Man*, Chap. V, pp. 219-220.

severe action of natural selection, the struggle for existence, is, in fact, attributable to a difference in the history of cultural experience—to nothing more, to nothing less.[87] As the *Unesco Statement by Experts on Race Problems,* issued to the world on July 18, 1950, puts it, "The scientific material available to us at the present time does not justify the conclusion that inherited genetic differences are a major factor in producing the differences between the cultures and cultural achievements of different peoples or groups. It does indicate, however, that the history of the cultural experience which each group has undergone is the major factor in explaining such differences. The one trait which above all others has been at a premium in the evolution of man's mental characters has been educability, plasticity. This is a trait which all human beings possess. It is indeed, a species character of *Homo sapiens.*" [88]

But the facts enabling us to arrive at such judgments were not available in Darwin's day in anything resembling the abundance of our own time, though had Darwin sought them as zealously as he sought facts to support his theory of struggle he would have found a sufficient number to make out a very good case for the essential equality of mankind. So far as mental qualities are concerned in the evolution of man the evidence strongly suggests that the premium

[87] For a fuller discussion of the facts see M. F. Ashley Montagu, *Man's Most Dangerous Myth: The Fallacy of Race,* 3d ed., Harper, New York, 1952.

[88] See Ashley Montagu, *Statement on Race,* Schuman, New York, 1951, p. 14.

has not been upon the development of special capacities but upon the development of the general capacity of plasticity or educability. What, under the action of natural selection, has been at a premium in the evolution of man is a kind of plasticity which seemingly would be favored in all human groups, namely, the ability to make rapid adjustments to rapidly changing conditions, to exhibit wisdom, maturity of judgment, and the ability to get along with other people. Natural selection would favor the development of such traits everywhere and in all societies. Such traits imply the ability to profit from previous experience, and the ability to apply it to new conditions. This has been the prime requirement for survival in all human groups, and not the development of this or that special ability. The effect of natural selection in all groups of man has probably been to render genetic differences in personality traits, as between individuals and particularly as between ethnic groups or races, relatively unimportant compared to their expressive plasticity.[89]

SUPERIOR RACES AND NATIONS

Darwin very definitely believed in the doctrine of superior and inferior races. The following passages from *The Descent of Man* may be quoted as illustrative of this fact. "The belief that there exists in man some close relation between the size of the brain and

---

[89] For further discussion see Th. Dobzhansky and M. F. Ashley Montagu, "Natural Selection and the Mental Capacities of Mankind," *Science,* vol. 105, 1947, pp. 587-590.

the development of the intellectual faculties is supported by the comparison of the skulls of savage and civilised races, of ancient and modern people, and by the analogy of the whole vertebrate series." [90]

The analogy, alas, limps, and as for the size of the brain or cranial capacity having anything to do, within the limits of the normal range, with intelligence, that is a myth which has long been exploded.[91] But this passage indicates something of the nature of Darwin's belief in the physical basis of the differences between the so-called "races."

On the differences in "progress" achieved by nations Darwin remarks that "it depends on an increase in the actual number of the population, on the number of the men endowed with high intellectual and moral faculties, as well as on their standard of excellence." [92] This is a rather simplified explanation of a complex problem. The history of the nation's experience, the opportunities for schooling and higher education, industrial development, kind of government, national wealth, and numerous similar factors remain unconsidered in Darwin's judgment. Yet we know these to be extremely important factors in determining the degree of a nation's progress.

---

[90] *The Descent of Man,* Chap. II, pp. 81-82.

[91] See the discussion in M. F. Ashley Montagu, *Man's Most Dangerous Myth: The Fallacy of Race,* 3d ed., Harper, New York, 1952, pp. 54 *et seq.;* also the same author's *An Introduction to Physical Anthropology,* 2d ed., C C Thomas, Springfield, Ill., 1951; Donald G. Paterson, *Physique and Intellect,* Century, New York, 1930.

[92] *The Descent of Man,* Chap. V, p. 216.

When Darwin uses the term "lower races" he quite clearly means "biologically inferior races." [93]

He approvingly quotes a Mr. Greg, to the effect that "Given a land originally peopled by a thousand Saxons and a thousand Celts—and in a dozen generations five-sixths of the population would be Celts, but five-sixths of the property, of the power, of the intellect, would belong to the one-sixth of Saxons that remained. In the eternal 'struggle for existence,' it would be the inferior and *less* favoured race that had prevailed—and prevailed by virtue not of its good qualities but of its faults." [94] The Irish, in short, would have conquered the Scots, not to mention the English.

That the hierarchical differences between the races of men owe their being to the action of natural selection is made quite clear in a letter which Darwin wrote toward the end of his life. This letter was addressed to William Graham (1839-1911).[95] The letter is dated July 3, 1881, and the passage to which I wish to make reference runs as follows, "Lastly, I could show fight on natural selection having done and doing more for the progress of civilization than you seem inclined to admit. Remember what risk the nations of Europe ran, not so many centuries ago of being overwhelmed by the Turks, and how ridiculous such an idea now is! The more civilized so-called Caucasian races have beaten the Turkish hollow in the struggle for existence. Looking to the world at no

---

[93] *Ibid.*, Chap. V, p. 212.

[94] *Ibid.*, p. 213.

[95] Author of a distinguished book, *The Creed of Science*, Kegan Paul, London, 1881.

very distant date, what an endless number of the lower races will have been eliminated by the higher civilized races throughout the world." [96]

MAN'S FUTURE AS SEEN BY DARWIN

This sounds like a biological jingo imperialist. What Darwin believed in with respect to man, he states quite clearly and unequivocally at the very end of *The Descent of Man*. Darwin writes, "Man, like every other animal, has no doubt advanced to his present high condition through a struggle for existence consequent on his rapid multiplication; and if he is to advance still higher, it is to be feared that he must remain subject to a severe struggle. Otherwise he would sink into indolence, and the more gifted men would not be more successful in the battle of life than the less gifted. Hence our natural rate of increase, though leading to many and obvious evils, must not be greatly diminished by any means. There should be open competition for all men; and the most able should not be prevented by laws or customs from succeeding best and rearing the largest number of offspring. Important as the struggle for existence has been and even still is, yet as far as the highest part of man's nature is concerned there are other agencies more important. For the moral qualities are advanced, either directly or indirectly, much more through the effects of habit, the reasoning powers,

[96] Francis Darwin (editor), *The Life and Letters of Charles Darwin*, John Murray, London, 1888, vol. 1, p. 316.

instruction, religion, &c., than through natural selection; though to this latter agency may be safely attributed the social instincts, which afforded the basis for the development of the moral sense." [97]

FIRST NATURAL SELECTION—THEN CULTURAL FACTORS

There it is. Man has developed to his present high estate through a struggle for existence. Thus, evolution comes about through "open competition." Furthermore, if he is to advance still "higher" he must remain subject to "a severe struggle," otherwise he will degenerate. To maintain the struggle we must be careful not to diminish greatly the natural rate of increase of population. The poor we must have always with us.[98] "There should be open competition for all men." But there are more important agencies at work than the struggle for existence in respect of man's moral

[97] *The Descent of Man*, Chap. XXI, pp. 945-946.

[98] "With savages, the weak in body or mind are soon eliminated; and those that survive commonly exhibit a vigorous state of health. We civilised men, on the other hand, do our utmost to check the process of elimination; we build asylums for the imbecile, the maimed, and the sick; we institute poor-laws; and our medical men exert their utmost skill to save the life of every one to the last moment." Charles Darwin, *The Descent of Man*, p. 205. The poor-laws, in the opinion of Townsend, Malthus, and Darwin, were a great error, since they were artificially calculated to preserve the poor. Darwin felt that this interfered with the action of natural selection. It was for the same reason that he was opposed to birth-control.

qualities. These agencies are largely of a cultural nature. But lest anyone should conclude from this that Darwin felt that man was his own maker, there is this in the final paragraph of *The Descent of Man,* "Man may be excused for feeling some pride at having risen, though not through his own exertions, to the very summit of the organic scale." (pp. 946-947.) Man, then, has not risen through his own exertions. Either Darwin contradicts himself or we misunderstand him. Whatever the case may be, there can be no wonder that the German Military Staff could make good use of Darwin in all good conscience, that Social Darwinists could enjoy a regular field day quoting him lock, stock, and barrel, while others imagined that they could assert with equal confidence that Darwin was not a Social Darwinist. I believe, however, that we can safely interpret Darwin to mean that natural selection, the struggle for existence, has been the supreme arbiter in determining the evolution of man's physical characters, as well as his "social instincts," but that once he arrived at a high stage of moral development, cultural factors assumed the supreme place in determining his social development. Before that high stage of development was reached, however, natural selection had been the dominant determinant of further development. The high stage of moral development once being reached, cultural factors take over the function of natural selection. Though Darwin is somewhat ambivalent in his attitude toward cultural factors as agencies in the evolution of man, it is reasonably clear that he considered man's future development to be dependent largely, if not entirely, upon the improvement in his morality through cultural means. It was not altogether Darwin's fault if subsequent generations have come to confuse the morality of evolution with the evolution of morality.

CO-OPERATION IN NONHUMAN AND HUMAN ORGANISMS

In spite of the fact that in the body of his book and in its penultimate paragraph Darwin prominently mentions the co-operative behavior of monkeys, he fails to recognize the importance of co-operation as a factor of evolution. It is clear, however, that Darwin, although failing to recognize co-operation as a factor of evolution in nonhuman organisms, did grant it a place in the evolution of man both in the past and in an increasing role in the future.[99]

Darwin failed to see the importance of co-operative behavior in the evolution of nonhuman organisms, but he did to some extent recognize that in man, as Haldane has put it, "in so far as it makes for the survival of one's descendants and near relations, altruistic behaviour is a kind of Darwinian fitness, and may be expected to spread as a result of natural selection." [100]

In this latter sense, essentially conceived in terms of the future development of man, Darwin was a co-operator. But quite clearly he was at the same time a

---

[99] See *The Descent of Man,* Chap. IV, pp. 187-188. This work, p. 81.

[100] J. B. S. Haldane, *The Causes of Evolution,* Longmans, New York, 1935, p. 131.

non-co-operator, the leading exponent of the struggle-for-survival school. If it is true that the appeal to Darwinian theory for support of a social system based on ruthless competition and exploitation of colonial peoples represents a misinterpretation of evolutionary theory, the conclusion seems inescapable that Darwin himself was the first to make this misinterpretation. The few passages in which Darwin mentions altruism, and co-operation come exclusively from *The Descent of Man*, where in a book of a thousand pages they are virtually crowded out by numerous statements that appear to stand in direct and unequivocal contradiction to them. Darwin was going to have the cake of natural selection, which he had baked, and eat it, too. The flavoring was mostly "struggle for existence," but here and there there was a sprinkling of the thinnest kind of co-operation. But the important point is that it was there, and it was Darwin's doing. It was, however, far too thin to make any more than a fleeting impression upon the intellectual taste-buds of those who concentrated their attention upon the consumption of the main body of the cake. Darwin was, in fact, a struggle-for-survivor, and only at times a weak co-operator. As Geoffrey West has put it in an admirable study of Darwin, "It is in relation to the Struggle for Existence, particularly, that Darwin and those who came immediately after him are felt, despite his reservations, to have scanned their evidence too narrowly and jumped to conclusions too hastily. They overstressed the intensity of the struggle, fixing attention upon biological competition and disregarding biological co-operation. Seeing some organisms as predatory, they ascribed such motives to all. Noting the great tusks and talons given to species extinct and living, they supposed it was by weight of these they held their places. They drew for humanity an ancestry in which the form of the ape was joined to the ferocity of the tiger. By tigerishness the ape had raised himself to man; by tigerishness would man raise himself to whatever higher form evolutionary progress held in store for him." [101]

Indeed, whatever Darwin may have intended in *The Descent of Man,* the message which the world has derived from his work is "Multiply, vary, let the strongest live and the weakest die!" In so far as the survival of the fittest is concerned in this process in the world in which he lived, Darwin took it to be equivalent to the survival of the richest—the middle classes and the wealthy. His own social background and the period in which he lived largely conditioned this view in him.[102] These factors also played a part

---

[101] Geoffrey West, *Charles Darwin: The Fragmentary Man,* Yale University Press, New Haven, Conn., 1937, p. 321.

[102] Justifying, we cannot help but suspect, his own position Darwin writes, ". . . the inheritance of property by itself is very far from an evil; for without the accumulation of capital the arts could not progress; and it is chiefly through their power that the civilised races have extended, and are now everywhere extending their range, so as to take the place of the lower races. Nor does the moderate accumulation of wealth interfere with the process of selection. When a poor man becomes moderately rich, his children enter trades or professions in which

in the interpretations which he gave to such things as the evolutionary status of races, the extermination of "lower" by "higher" races, and the struggle for life being most severe between individuals and varieties of the same species. We have seen how highly questionable some of Darwin's interpretations are, and it would be possible to go on and show how faulty were so many of his other interpretations, but since this is a task which has already been adequately performed in many of the critical works mentioned in this essay, it will be unnecessary to prolong its length by discussing them further here.

And so we come to the end of our journey with, I hope, undiminished respect for Darwin and greater understanding of his work. Alonzo Myers has remarked that "As a scientist Darwin ranks with Galileo and Newton; as a philosopher he ranks somewhere between Herbert Spencer and Gobineau. Unfortunately, however, his highly questionable system of values derives an unmerited prestige from the acknowledged value of his scientific findings." [103] I believe that this is a fair judgment.

---

there is struggle enough, so that the able in body and mind succeed best. The presence of a body of well-instructed men, who have not to labour for their daily bread, is important to a degree which cannot be over-estimated; as all high intellectual work is carried on by them, and on such work, material progress of all kinds mainly depends, not to mention other and higher advantages." *The Descent of Man*, Chap. V, pp. 207-208.

[103] Henry Alonzo Myers, *Are Men Equal?* Putnam, New York, 1945, p. 110.

THE MISSING PIECE IN DARWIN'S PUZZLE

Darwin bequeathed to the world a fragmentary, a partial, an incomplete truth. It helped to dispel abstract and spiritual fears by supplying facts and theories with which one could grapple, only to lead to a world in which fear has become endemic upon a scale hitherto unknown, with its locus not in the abstract but in the real world. Darwin helped to establish such seeming paradoxes as that good could flow from evil, and that in the biological sense such evils were really good, and so, "from the war of nature, from famine and death, the most exalted object which we are capable of conceiving, namely, the production of the higher animals, directly follows." Darwin's seeming paradoxes have long been the platitudes of the day. The doctrine of the survival of the fittest, the struggle for existence, natural selection, competition—all these are the same things—is the religion of contemporary man. Whatever other religion he may profess and pay lip service to constitutes but the smoke of incense burnt before an empty shrine. This is our inheritance from such genuinely decent and noble men as Darwin, Huxley, Haeckel, and Spencer. Let us honor them for what they were, and let us avoid making them the scapegoats for our own belatedly recognized deficiencies. The work and the writings of those who saw what was fallacious in the ideas of these men have been long with us, but they have received little attention. The thinkers who were capable of liberating us from the errors of Darwinism had, if anything, vastly more respect for the teachings of the Darwinians than those who uncritically accepted those teachings. The critics, at least, had studied them critically.

Darwin was the great liberator—from one prison into another. He liberated the thought of man from the shackles of an anthropomorphic theology, and cast him into the freedom of a crass mechanical materialism which leads straight to the freedom of destruction—and this, largely because of the out-of-focus, dangerously incomplete view of the nature of life which he presented. What a difference it would have made to the world had Darwin been able to write *The Origin of Species* and *The Descent of Man* with the emphasis placed on co-operation rather than upon struggle! The missing piece, in the puzzle which Darwin thought he had successfully put together, is the principle of co-operation. The next fifty years will, I believe, see that principle firmly established in scientific and in secular thought. Let us, however, always remember that we must avoid making, as T. H. Huxley said, the heresies of one generation the superstitions of another.

2

THE CONTEMPORARY THEORY
OF NATURAL SELECTION

Having devoted so much space to showing the reader what was wrong with Darwin's conception of natural selection, it may perhaps be useful to present here a brief account of natural selection as it is understood by contemporary biologists.

Changes in the character of the hereditary particles, the genes of the living cells, are called mutations. Mutations provide the raw materials of evolution. Such changes or variations in the hereditary particles occur in individual organisms and are hereditarily transmitted to their offspring. In a population in which there exist differences in the way in which such mutant genes are distributed, that is to say, in which some types or groups possess such mutant genes and some do not, *if for any reason one type or group leaves a greater surviving progeny than others,* certain genes and associations of genes will become more, and others less, frequent in succeeding generations. This is what is meant by natural selection—differential fertility, resulting in the perpetuation of new genotypes.

Organisms adapt themselves to their environment. Natural selection is the principle that attempts to account for this adaptation. Progressive adaptations are believed to be the steps through which evolution takes place. The essence of the concept of natural selection is differential fertility, that is, the carriers of different complexes of genes, genotypes, contribute in different proportions to the gene pool of succeeding generations. The (Darwinian) fitness or *adaptive value* of the genotype is the relative capacity of the carriers of a given genotype (complex of genes carried by each individual organism) to transmit their genes to the gene pool of succeeding generations. Adaptive value or fitness is a statistical conception that sums up the reproductive efficiency of a genotype in a particular environment. It should be clear that the adaptive value is influenced by the ability of the type to survive. Furthermore, there are many variables that determine adaptive value (reproductive efficiency), such as fecundity (the physiological capacity to participate in reproduction), sexual drive, fertility (fecundity expressed in performance), the duration of the reproductive period. Clearly, the degree of sociality, of co-operativeness, prevailing in any group will constitute an important factor in influencing its fertility rate as well as its survival rate. It cannot be too often repeated that all animals are members of social groups, and that there is no such thing as a solitary animal, even though some groups are much less integrated than others. Social life is, among other things, a means of ensuring reproduction. To the extent, therefore, that any group is less social, less fully integrated, than another, it is likely to be differentially less fertile. All other factors being equal, with the exception of the degree of social integration, it should be obvious that the group in which its members are closely integrated, are often together, is the group that is likely to leave a greater surviving progeny than the group whose members are less integratedly social. Viability, reproductive capacity, and the like may be perfect, but if fertility is lower in one type or group than in another, the latter, in the same environment, will have an increasingly higher adaptive value than the former. Indeed, a higher fertility rate, from whatever causes arising, may compensate for deficient vigor or other deficiencies. And this is the important point: natural selection favors those carriers of genotypes in variable groups or populations that, from whatever causes arising, leave a greater progeny than do, in the same environment, carriers of other genotypes in such variable groups or populations.

It is important to note that no mention is made of "competition" or "struggle" in this account of natural selection. The abuse and misuse of these terms has already been discussed in this essay, and the desirability of avoiding their usage in any discussion of the causes of evolution has been emphasized.[104] Nevertheless, competition, struggle, as well as co-operation,

---

[104] See pages 43-72, this essay.

occur in nature. Combative competition and struggle *may* in some cases lead to one type leaving a greater progeny than another, but it does not necessarily follow that such processes will always do so. Furthermore, combative competition and struggle are not by any means necessary parts of the modern theory of natural selection, though such processes as possible determinants of differential fertility must always be taken into consideration in the analysis of the factors leading to fitness or adaptive value.

Recognition of the adaptive value of co-operation as a factor in evolution clarifies and enlarges the conception of natural selection, but by no means replaces it. Dobzhansky has stated the situation perfectly. He writes, "Cooperation, competition, struggle, and various combinations of these forms of relationships between organisms may, under various circumstances and at different times, augment or reduce the adaptive values of carriers of certain genotypes. Whether cooperation or disoperation prevails at a given time in a given environment depends upon the adaptive exigencies of the situation." [105]

There can be little doubt that the adaptive exigencies of man's situation in the world today make it abundantly clear that if he is to survive he must cease to compete disoperatively and begin, with greater awareness than he has in the past, to co-operate.

One more point: Dobzhansky writes that "In social organisms the survival and vigor of an individual is

---

[105] Th. Dobzhansky, *Genetics and the Origin of Species,* 3d ed., Columbia University Press, New York, 1951, p. 79.

adaptively important only if it contributes to the perpetuation of the colony or the group." [106] This is true, but the exact converse of this is also true, namely, that in organisms the survival and vigor of a colony or a group is adaptively important insofar as it contributes to the perpetuation of the individual.[107] With respect to man the latter is by far the more significant way of phrasing the truth than is the former, although both are true, and both are necessarily complementary.

---

[106] *Ibid.*, p. 78.

[107] For a brief development of this point see Ashley Montagu, *On Being Human*, Schuman, New York, 1950.

3

SOME REPRESENTATIVE STATEMENTS OF "THE STRUGGLE FOR EXISTENCE" SCHOOL OF THOUGHT

In this Appendix are listed some representative statements by thinkers belonging to "the struggle for existence" school of thought. General reference is made to these in the first paragraph on page 9 of this volume.

The following passage is from the pen of Ernst Haeckel (1834-1919). Distinguished zoologist and prolific writer, he did more than anyone else to popularize the "nature red in tooth and claw" conception of life in Germany. The passage is typical of many such statements made by scientists and others during the nineteenth century.

"The theory of selection teaches that in human life, as in animal and plant life, everywhere and at all times, only a small and chosen minority can exist and flourish, while the enormous majority starve and perish miserably and more or less prematurely. . . . The cruel and merciless struggle for existence which rages throughout living nature, and in the course of nature *must* rage, this unceasing and inexorable competition of all living creatures is an incontestable fact; only the picked minority of the qualified fittest is in a position to resist it successfully, while the great majority of the competitors must necessarily perish miserably. We may profoundly lament this tragical state of things, but we can neither controvert nor alter it. 'Many are called, but few are chosen.' This principle of selection is as far as possible from democratic; on the contrary it is aristocratic in the strictest sense of the word." Ernst Haeckel, *Freedom in Science and Teaching*, Appleton, New York, 1879, p. 93.

Here is Darwin on the relations between the races: "At some future period, not very distant as measured by centuries, the civilised races of man will almost certainly exterminate, and replace, the savage races throughout the world." Charles Darwin, *The Descent of Man*, 2d ed., 1874, pp. 241-242.

Herbert Spencer (1820-1903), English sociologist and philosopher, in his widely read *Principles of Sociology* (issued 1876-1880), wrote: "Not simply do we see that in the competition among individuals of the same kind, survival of the fittest has from the beginning furthered production of a higher type; but we see that to the unceasing warfare between species is mainly due both growth and organization. Without universal conflict there would have been no development of the active powers." Herbert Spencer, *Principles of Sociology*, vol. 2, Appleton, New York, 1876, pp. 240-241.

William Graham Sumner (1840-1910), Professor of Political and Social Science at Yale, was for many years prominently occupied convincing the public that Darwinian ideas lent support to the *status quo*. In a lecture delivered in 1879 Sumner declared that "Many economists seem to be terrified that distress and misery still remain on earth and promise to remain as long as the vices of human nature remain. Many of them are frightened at liberty, especially under the form of competition, which they elevate into a bugbear. They think it bears harshly on the weak. They do not perceive that here 'the strong' and 'the weak' are terms which admit of no definition unless they are made equivalent to the industrious and the idle, the frugal and the extravagant. They do not perceive, furthermore, that if we do not like the survival of the fittest, we have only one possible alternative, and that is the survival of the unfittest. The former is the law of civilization; the latter is the law of anti-civilization. We have our choice between the two, or we can go on, as in the past, vacillating between the two, but a third plan —the socialist desideratum—a plan for nourishing the un-fittest and yet advancing in civilization, no man will ever find." William Graham Sumner, *The Challenge of the Facts, and Other Essays*, Yale University Press, New Haven, Conn., 1914, p. 68. "Our civilization," says Sumner, "ordinarily veils from us the fact that we are rivals and enemies to each other in the competition of life."

Karl Pearson (1857-1936), distinguished English bio-metrician and social biologist, wrote, at the beginning of the century, in a book which is now a classic of science, as follows: "It is a false view of human solidarity, a weak humanitarianism, not a true humanism, which regrets that a capable and stalwart race of white men should replace a dark-skinned tribe which can neither utilize its land for the full benefit of mankind, nor contribute its quota to the common stock of human knowledge." "This sentence," adds Pearson, "must not be taken to justify a brutalizing destruction of human life. The anti-social effects of accelerating the survival of the fittest may go far to destroy the preponderating fitness of the survivor. At the same time there is cause for human satisfaction in the replacement of the aborigines throughout America

and Australia by white races of far higher civilization." [108]
Alice James—sister of William and Henry—wittily remarked upon "the profound and irreconcilable . . . con-
viction" of the English "that outlying regions are their preserves, for they alone of human races massacre savages
out of pure virtue." *Alice James, Her Brothers—Her Journal,* edited by A. R. Burr, Dodd, New York, 1934, p. 138.

Theodore Roosevelt (1858-1919), twenty-fifth president of the United States, wrote, in 1899, "We cannot
avoid the responsibilities that confront us in Hawaii,
Cuba, Porto Rico, and the Philippines. All we can decide
is whether we shall meet them in a way that will redound
to the national credit, or whether we shall make of our
dealings with these new problems a dark and shameful
page in our history. . . . The timid man, the lazy man,
the man who distrusts his country, the over-civilized man,
who has lost the great fighting, masterful virtues, the ignorant man, and the man of dull mind, whose soul is
incapable of feeling the mighty lift that thrills 'stern men
with empires in their brains'—all these, of course, shrink
from seeing the nation undertake its new duties. . . .

"I preach to you, then, my countrymen, that our country calls not for the life of ease but for the life of strenuous
endeavor. The twentieth century looms before us big with
the fate of many nations. If we stand idly by, if we seek
merely swollen, slothful ease and ignoble peace, if we
shrink from the hard contests where men must win at
hazard of their lives and at the risk of all they hold dear,
then the bolder and stronger peoples will pass us by, and
will win for themselves the domination of the world."
Theodore Roosevelt, "The Strenuous Life," *The Works of*

[108] Karl Pearson, *The Grammar of Science,* 2d ed., Black,
London, 1900; revised reprint, Everyman's Library, Dutton, New York, 1937, p. 310.

Theodore Roosevelt, Scribner, New York, vol. 13, pp. 322-323, 331.

General Friedrich von Bernhardi (1849-1930), German cavalry general and author of the famous *Germany and the Next War* (1912), asserted that "War is not merely a necessary element in the life of nations but an indispensable factor of culture, in which a truly civilized nation finds the highest expression of strength and vitality. . . . War gives a biologically just decision, since its decisions rest on the very nature of things. . . . It is not only a biological law, but a moral obligation, and, as such, an indispensable factor in civilization." General Friedrich von Bernhardi, *Germany and the Next War,* Longmans, New York, 1912, pp. 16-37.

An American equivalent of von Bernhardi was General Homer Lea (1876-1912), of the Chinese army, who wrote that "As physical vigor represents the strength of man in his struggle for existence, in the same sense military vigor constitutes the strength of nations; ideals, laws and constitutions are but temporary effulgences, and are existent only so long as this strength remains vital. As manhood marks the height of physical vigor among mankind, so the militant successes of a nation mark the zenith of its physical greatness." General Homer Lea, *The Valor of Ignorance,* Harper, New York, 1909, pp. 8, 11.

Sir Arthur Keith (b. 1866), well-known English anatomist and physical anthropologist, in a famous address, said, "Without competition Mankind can never progress; the price of progress is competition. Nay, race prejudice and, what is the same thing, national antagonism, have to be purchased, not with gold, but with life. Nature throughout the past has demanded that a people who seeks independence as well as peace can obtain these privileges only in one way—by being prepared to sacrifice their blood to secure them. Nature keeps her human orchard healthy by pruning; war is her pruning-hook. We cannot dispense with her services." Sir Arthur Keith, *The Place of Prejudice in Civilization,* Day, New York, 1931, p. 50.

Lord Elton, writing in 1942, states, "War, however much we hate it, is still the supreme agent of the evolutionary process. Blind, brutal and destructive, it remains the final arbiter, the one test mankind has yet contrived of a nation's fitness to survive." Lord Elton, *Saint George and the Dragon,* Collins, London, 1942.

"War is the most natural, the most every-day matter. War is eternal, war is universal. There is no beginning and there is no peace. Any struggle is war." Adolf Hitler (1889-1945?), Austrian-German chancellor of Germany and president of the Nazi Party, as quoted by Hermann Rauschning in *The Voice of Destruction,* Putnam, New York, 1940, pp. 7-8.

"You can't make the world all planned and soft. The strongest and best survive—that's the law of nature after all—always has been and always will be." Middletown business man, quoted by Robert S. and Helen M. Lynd, *Middletown in Transition,* Harcourt, New York, 1937, p. 500.

"You know, my business isn't distributing roofing—its principally keeping my competitors from distributing roofing. Same with you. All we do is cut each other's throats and make the public pay for it!"

"Look here, now, Paul! You're pretty near darn talking socialism!" exclaims Babbitt.

And Paul answers, "Oh yes, of course I don't really exactly mean that—I s'pose. Course—competition—brings out the best—survival of the fittest—but—" Sinclair Lewis, *Babbitt,* Harcourt, New York, 1922; Bantam Books, p. 65.

"The process commonly called 'the survival of the fittest' ceaselessly takes place in the business world in a competitive economic society like ours.

"Through such natural selection the four-flushers and second-guessers of business are soon eliminated. They cannot find 'angels' permanently to fund their deficits." From a "Statement on the American system of free enterprise which makes possible such institutions for service as Armour and Company," signed by the president of the company, G. A. Eastwood, and appearing as a full-page advertisement on the inside of the front cover of the *Woman's Home Companion,* September, 1944.

4

COMPETITION, COOPERATION, AND THE UNCONSCIOUS

A psychoanalytic interpretation of the unconscious motivation of many who believe in competition as a way of life—as well as many who believe in cooperation—has recently been given by Dr. R. E. Money-Kyrle. This should be of great interest to readers of the present volume. Discussing the problem of inter-group relations Dr. Money-Kyrle writes:

"Among the various views expressed on this problem, two stand out in glaring contrast with each other. According to the first, inter-group relations must always be ruthlessly competitive. This view is often assumed to be an inevitable consequence of the Darwinian theory applied to group selection. If indeed only the strongest, the most skilful and most aggressive groups can survive in the struggle for existence, selection must favour the development of groups that are ever more cunning and ferocious in their dealings with their fellows. Some people who argue in this way accept their own conclusions with regret, as a bitter truth which they would escape from if they could. But others seem rather to flaunt it with triumph as if to justify their personal admiration for whatever is strong, or ruthless, and their contempt for humanistic scruples which they equate with weakness. According to the opposite view, the natural state of man is one of universal peace and brotherhood. Those who believe this differ in their opinions whether such a state ever existed in the past, and, if so, about whether it was lost through the sins of Adam or the greed of capitalists. They also differ about how it is to be gained, whether by an inner change of heart or by some external revolution. And those who believe it can be gained by social changes differ about the kind of changes required. But they all agree that the aim they share in common can somehow be attained.

"In the light of our understanding of unconscious guilt and of the various defences against it, we can see one reason why so many people should wish to hold one or other of these opposite views on the future of group relations. For, if their own conscious or unconscious predatory impulses arouse strong feelings of unconscious guilt, they will tend to deny either the guilt or the impulses. Those who exalt the inevitability of conflict deny their sense of guilt and develop a defensive pride in their freedom from all scruples. Those who cling to a vision of a world without strife, or even competition, deny at least some part of the predatory aggression that threatens to disturb their relations with their fellows. And they usually project it on to scape goats who, at the same time, provide them with an excuse and an object for it." [109]

---

[109] R. E. Money-Kyrle, *Psychoanalysis and Politics*. Duckworth, London, 1951, New York, Norton, 1952, pp. 91-92.

5

ANNOTATED BIBLIOGRAPHY

The books and other writings listed below do not include all those referred to in the text, and include a number not there mentioned. The principal purpose of this bibliography is to provide a ready answer to the question: "Where can I read further on this subject?" Many books are here listed which deserve to be far better known than they are at present.

ALLEE, W. C. *Animal Aggregations.* University of Chicago Press, Chicago, 1931.

The standard and the best scientific work on association and co-operation among animals. Facts as well as theories are examined, and certain basic sociological principles formulated.

————*Animal Life and Social Progress.* Williams & Wilkins, Baltimore, Md., 1932.

A popularly written account of the nature of social life and its significance for the social group.

————*Cooperation Among Animals.* Schuman, New York, 1951.

First published in 1938 as *The Social Life of Animals,* this admirable book brings up to date our knowledge of the nature of the social or co-operative processes that characterize animals and animal groups—including man. It is the best and most readable work of its kind to place in the hands of anyone seeking a ready acquaintance with the facts.

————EMERSON, A. E., PARK, O., PARK, T., and SCHMIDT, K. P. *Principles of Animal Ecology.* Saunders, Philadelphia, 1949.

The best and most authoritative work on the relations of animals to their environments—illustrating the importance of the principle of harmonic adjustment. A very big book in all senses of the word.

BAGEHOT, W. *Physics and Politics, or Thoughts on the Application of the Principles of Natural Selection and Inheritance to Political Society.* Appleton, New York, 1872.

This brilliant work exercised a considerable influence upon economists. Bagehot not only applied the principle of natural selection to the development of peoples but also saw the importance of co-operation as a factor in their development.

BARNES, H. E., and BECKER, H. *Social Thought From Lore to Science.* 2 vols. Heath, Boston, 1938.

Pages 693 to 742 of volume I deal very illuminatingly with Social Darwinism, while the preceding chapter sets the background, and the one following traces the fall, of evolutionism in sociology.

BARZUN, J. *Darwin, Marx, Wagner*. Little, Brown, Boston, 1941.

>An excellent volume, very readable, and especially good on Darwin and Darwinism.

BATES, M. *The Nature of Natural History*. Scribner, New York, 1950.

>A valuable discussion of the nature of nature from the field naturalist's viewpoint. A welcome restorative to the facts and an illuminating release from the laboratory biologist's conception of nature. Here the co-operative processes receive their due.

————*Where Winter Never Comes*. Scribner, New York, 1952.

>A delightful account of the tropics as they really are, and not as desk naturalists have stated them to be. A valuable corrective to the popular conception of the "jungle."

BERG, L. S. *Nomogenesis, or Evolution Determined by Law*. Constable, London, 1926.

>A distinguished contribution to evolution by a Russian biologist, in which co-operation is given proper emphasis as a factor in the process of evolution.

BERNARD, H. M. *Some Neglected Factors in Evolution*. Putnam, New York, 1911.

>An important and much "neglected" book in which the principle of co-operation is brought into true focus as a factor in evolution.

BERNHARDI, F. VON. *Germany and the Next War*. Longmans, New York, 1914.

>Published in Germany in 1912 and in England

and the United States in 1914, this best-seller frankly states the warlike intentions of the German militarists, and rests its case firmly on the thesis that war is the best test of a nation's fitness to rule. Darwin figures prominently as an authority giving scientific validation to such notions—as, indeed, he does.

BOUGLE, C. "Darwinism and Sociology," in SEWARD, A. C. (editor), *Darwin and Modern Science.* Cambridge University Press, New York, 1909, pp. 465-476.
An excellent discussion.

BROOKS, L. M. "Fifty Years' Quest for Social Control," *Social Forces*, vol. 29, 1950, pp. 1-8.
"As we start on the second fifty years of the century, still questing for the social control that will come only through the cooperative process, there is the challenge to all scientists and to all mankind to work as never before." "The challenge for us all . . . is to cooperate or perish."

BÜCHNER, L. *Aus dem Geistesleben der Thiere.* 2d ed., Berlin, 1877.

———*Liebe und Liebesleben in der Thierwelt*, Berlin, 1879.
In these two works the author consistently emphasizes the factor of co-operation in the relations between animals and their relations to their environment.

CANTRIL, H. "Toward a Scientific Morality," *Journal of Psychology*, vol. 27, 1949, pp. 363-376.
An important paper by a distinguished psychologist on the scientific bases for ethics.

CASSON, S. *Progress and Catastrophe*. Hamish Hamilton, London, 1937.
> An anatomy of the human adventure.

CASTRO, J. DE. *The Geography of Hunger*. Little, Brown, Boston, 1952.
> A most effective reply to the Malthusians.

CHASE, STUART. *The Proper Study of Mankind*. Harper, New York, 1948.
> Subtitled "An Inquiry Into the Science of Human Relations," Chase's book soundly and readably presents a comprehensive account of the findings of contemporary social scientists concerning the nature of human behavior.

———*Roads to Agreement*. Harper, New York, 1951.
> An admirable account of the findings of modern scientists concerning the improvement of the ability of people to get along with one another.

CHILDE, V. G. *History*. Cobbett Press, London, 1947.

———*Man Makes Himself*. Mentor Books, New York, 1951.

———*What Happened in History*. Mentor Books, New York, 1946.
> Three books by the greatest living prehistorian, indispensable for an understanding of how man developed to his present condition.

COOLEY, C. H. *Human Nature and the Social Order*. Scribner, New York, 1902.
> In this volume the great American sociologist reduced to absurdity the notion that self-reliant individualism was the ultimate achievement of

evolution. His discussion of human nature is still very much worth reading, even though it is now half a century old.

CORWIN, E. S. "The Impact of the Idea of Evolution on the American Political and Constitutional Tradition," in PERSONS, S. (editor), *Evolutionary Thought in America.* Princeton University Press, Princeton, N. J., 1950, pp. 182-199.

A brief and excellent study.

DARWIN, C. *The Descent of Man.* John Murray, London, 1871.

The second edition was published in 1874. This is the work in which Darwin speaks of the "social instincts" of man, thus recognizing his co-operative drives, but it is also the work in which Darwin displays the fact that he is a Social Darwinist. This is indispensable reading.

————*On the Origin of Species by Means of Natural Selection, Or the Preservation of Favoured Races in the Struggle for Life.* John Murray, London, 1859.

This is the most influential book ever published in the whole history of biology. The first edition is listed above, the sixth and last edition was published in 1872. The latter has been many times reprinted, most recently by the Philosophical Library, New York, 1951.

DARWIN, F. (editor). *Charles Darwin's Autobiography.* Schuman, New York, 1950.

A condensed version of *The Life and Letters of Charles Darwin,* 3 vols., John Murray, London, 1888. Indispensable for any genuine understanding of Darwin the man and thinker.

DOBZHANSKY, TH. *Genetics and the Origin of Species.* 3d ed.,
Columbia University Press, New York, 1951.
   The contemporary equivalent of Darwin's *Origin
of Species.* A fundamental and indispensable book.

DRACHMAN, J. M. *Studies in the Literature of Natural
Science.* Macmillan, 1930.
   An admirable study of the Darwinians and others.

DRUMMOND, HENRY. *The Ascent of Man.* Hodder & Stoughton, London, 1894.
   The Lowell Lectures delivered at Boston. This
is one of the great books of our culture in which
Drummond foreshadows most of the great discov-
eries that have since been made in the fields of the
biological and social sciences concerning the nature of nature and of human nature. A must book.

DURBIN, E. F. M., AND BOWLBY, J. *Personal Aggressiveness
and War.* Columbia University Press, New York, 1939.
   Important because it traces the development of
co-operation and aggression in children, and shows
that co-operation is a natural process like growth,
which if interfered with leads to aggression and
fighting. The ease with which men fall back upon
violence and armaments to solve their adult problems is related to their childhood conditioning.

EMERSON, A. E. "The Biological Basis of Social Cooperation," *Transactions Illinois Academy of Science,* vol. 39,
1946, pp. 9-18.
   A splendid account of the facts. A fundamental
paper.

ESPINAS, A. V. *Des Sociétés animales.* Librairie Baillière,
Paris, 1878.
   A remarkable work on the social life of animals,

the third edition of which appeared in 1924. The author, an entomologist, emphasized the factor of co-operation in evolution and in the survival of animal groups. He was too much ahead of his time and, hence, little attention was paid to his work.

FAIRCHILD, H. N. *The Noble Savage*. Columbia University Press, New York, 1928.

A study in romantic naturalism in which the development of the idea of the "noble savage" in English literature is traced from the earliest days to the nineteenth century. Delightfully readable.

FARIS, R. E. L. "Evolution and American Sociology," in PERSONS, S. (editor), *Evolutionary Thought in America*. Princeton University Press, Princeton, N. J., 1950, pp. 161-180.

A short but informative study of Social Darwin·ism and its effect upon American sociologists.

FISKE, E. F. *The Letters of John Fiske*. Macmillan, New York, 1940.

There are few things that can convey better the enthusiasm, the almost religious fervor, with which Darwinism and Spencerism were embraced by some of the best minds of the day in the United States, than these letters of one of the molders of American thought who met most of the principals involved. An enchanting and important book.

FRANK, LAWRENCE K. *Nature and Human Nature*. Rutgers University Press, New Brunswick, N. J., 1951.

A brilliant book on human nature and human relations, by one of the great leaders of work and of thought in this area of knowledge of our day.

————*Society as the Patient*. Rutgers University Press, New Brunswick, N. J., 1948.

Thirty great essays on the human situation.

GEDDES, P. "Biology," in *Chambers Encyclopaedia*, London, Edinburgh, and Philadelphia, 1882.

In his articles on "Biology" and "Evolution" in this grand encyclopaedia Patrick Geddes briefly made what were probably the first corrective comments on "competitive" Darwinism offered by an English biologist.

————and THOMSON, J. A. *The Evolution of Sex*. Scott, London and New York, 1889, Scribner & Welford, New York, 1890.

One of the earliest English books in which the importance of co-operation as a factor of evolution was emphasized. This occurs chiefly in the last few chapters.

GESSLER, C. *The Reasonable Life*. Day, New York, 1950.

By an American newspaper man about the way of life of Polynesian islanders, and what it means to be without the tension, hurry, and worry of western man, and without his kind of competition.

GIBSON, R. W. *The Morality of Nature*. Putnam, New York, 1923.

A book which deserves to rank as one of the great works of our century—yet one to which I have never seen a reference. It is a scientifically sound and esthetically beautiful exposition of the principle of co-operation. There are riches enough in this volume to supply generations of workers and thinkers with ideas.

HAMMOND, J. L., and B. *The Bleak Age.* Pelican Books, London and New York, 1947.

A vivid and authoritative account of life in nineteenth century England in the throes of the industrial revolution, the period during which Darwin lived and wrote.

HASKINS, C. P. *Of Societies and Men.* W. W. Norton, New York, 1951.

A most stimulating book.

HENKIN, L. J. *Darwinism in the English Novel, 1860-1910.* Corporate Press, New York, 1940.

A most attractive study of the impact of Darwinism upon Victorian fiction.

HIRST, E. W. *Ethical Love.* Allen & Unwin, London, 1928.

The author ably argues the case that ethical love has its roots in the fundamentally social nature of life and of the universe. A very stimulating book.

HOFSTADTER, RICHARD. *Social Darwinism in American Thought, 1860-1915.* University of Pennsylvania Press, Philadelphia, 1944.

This is by far the best and most brilliant study of its kind. Delightfully written, it traces the development of Social Darwinism in America with great skill.

HOLMES, S. J. *Life and Morals.* Macmillan, New York, 1948.

An important book by a distinguished biologist who really gets down to grips with the questions: What bearing have the facts of biology upon our understanding of men living in societies? What ethical considerations develop from such an understanding of the facts?

HUXLEY, J. S. *Evolution: The Modern Synthesis.* Harper, New York, 1942.

The best and most complete exposition of the modern theory of evolution.

HUXLEY, T. H. "The Struggle for Existence: A Programme," *The Nineteenth Century* (London), vol. 23, February, 1888, pp. 161-180.

It was in this famous essay that Huxley described the animal world as on the level of a gladiator's show, evoking from Kropotkin his famous reply in a series of articles in the same journal, afterward published as the book *Mutual Aid* (1902).

————and JULIAN S. *Touchstone for Ethics.* Harper, New York, 1947.

Grandfather and grandson discuss the relation of evolutionary theory to ethics—at a remove of fifty years. The Romanes Lecture of 1893 by "T.H." and that of 1943 by "J.S." are reprinted, together with much additional valuable material. This is a most important work, the most significant parts of which are contributed by J. S. Huxley.

JENSEN, J. V. *The Long Journey.* Knopf, New York, 1945.

An epic novel of the ascent of man, for which Jensen was awarded the Nobel Prize in 1944. Based on a knowledge of modern anthropology and archeology, the theme that runs through this poetic work is the rise of man through co-operation.

KAHLER, E. *Man the Measure.* Pantheon Books, New York, 1943.

This is the best history of the rise of man and of humanity I know. Written by a great scholar, the book has philosophic breadth as well as the freshness of an entirely new approach to history.

KESSLER. "Mutual Aid as a Law of Nature and the Chief Factor of Evolution," *Memoirs (Trudy) of the St. Petersburg Society of Naturalists,* vol. 9, 1880.

I have never seen this article. It was delivered as a lecture by Professor Kessler before the St. Petersburg Society of Naturalists on January 8, 1880, and was noticed in the English scientific journal *Nature* (London) January 21, 1880. This was undoubtedly due to Kropotkin, who wrote many pieces for *Nature.* The original is, of course, in Russian. It would be very nice to have a translation of it.

KLUCKHOHN, C. "The Way of Life," in *The Kenyon Review,* 1941, pp. 160-179.

The best short analysis of American culture I know.

KNOWLES, L. C. A. *The Industrial and Commercial Revolutions in Great Britain During the Nineteenth Century.* Routledge, London, 1947.

A solid account—necessary background material for the understanding of the environment in which Darwin flourished.

KORZYBSKI, ALFRED. *Manhood of Humanity.* Institute of General Semantics, Lakeville, Conn., 2d ed., 1950.

This is a great book, much underestimated, the theme of which, in Korzybski's own words, is that "ethics must conform to the *natural* laws of human *nature.*" But there are a great many other secondary themes, particularly that of "human engineering" and the concept of man as a "time-binding" creature.

KROPOTKIN, P. *Ethics: Origin and Development.* Tudor, New York, 1947.

> A continuation of *Mutual Aid* (see below). In this work Kropotkin applies the principle of mutuality, of co-operation, to men living in societies. Though left unfinished at his death in 1922, the volume is complete in itself. It is only Kropotkin's final conclusions which are lacking in a formal chapter.

————*Mutual Aid.* Penguin Books, London, 1939.

> This is the *great* and classic book in the field of evolution in which co-operation is introduced as a factor meriting at least as much attention as competition. First issued as a book in 1902, it has been several times reissued. The latest publisher is given above.

LACK, D. *Darwin's Finches.* Cambridge University Press, New York, 1947.

> A delightful study, providing a scientific demonstration of evolution without competition.

LANESSAN, J. M. A. "La Lutte pour l'Existence et l'Association pour la Lutte," *Bibliothèque Biologique,* Paris, 1881.

> With the emphasis on co-operation.

LATHAM, R. *In Quest of Civilization.* Jarrolds, London, 1946.

> One of the best and most readable of all books on the history of civilization.

LILLIE, R. S. *General Biology and Philosophy of Organism.* University of Chicago Press, Chicago, 1945.

> A most stimulating book by a great physiologist, particularly valuable for its twelfth chapter on the

129

special role of the psychic factor in living organisms, and the author's views on the nature of goodness and co-operation.

MC CULLOCH, J. R. *A Selection of Scarce and Valuable Economical Tracts, from the Originals of Defoe, Elking, Franklin, Turgot, Anderson, Schomberg, Townsend, Burke, Bell, and Others.* London, 1859.

A valuable collection of particular interest to us because it is the most accessible source for Townsend's *Dissertation on the Poor Laws,* in which both Malthus and Darwin are foreshadowed. See Townsend below.

MACFARLANE, J. M. *The Causes and Course of Organic Evolution.* Macmillan, New York, 1918.

A very original book by an academic biologist unafraid of admitting that he has been greatly influenced by Kropotkin. See especially the last four chapters.

MALTHUS, T. R. *An Essay on the Principle of Population, as It Affects the Future Improvement of Society, with Remarks on the Speculations of Mr. Godwin, M. Condorcet, and Other Writers.* J. Johnson, London, 1798.

The most influential book on population ever published. The fifth edition was published in 1817. There have been many editions since, and all of them are now out of print. The best edition is by James Bonar, Everyman's Library, Dutton, New York, 1926.

MASLOW, A. H. "Our Maligned Animal Nature," *Journal of Psychology,* vol. 28, 1949, pp. 273-278.

On the nature of human nature, its essential co-operativeness, and essential nonaggressiveness.

MEAD, M. (editor). *Cooperation and Competition Among Primitive Peoples.* McGraw-Hill, New York, 1937.
> The cultures of thirteen nonliterate peoples are examined for their co-operative and competitive traits. There is an excellent interpretative statement by the editor.

MILLER, E. M. "A Look at the Anatomy and Physiology of Groups," *Bios,* vol. 2, 1949, pp. 24-31.
> An admirable paper in which the principle of cc operation is most effectively discussed.

MILLER, HUGH. *The Community of Man.* Macmillan, New York, 1949.
> A philosopher rejects traditional philosophy and develops a philosophy on the basis of biological concepts, with the emphasis on co-operation. Indispensable reading.

MITCHELL, P. C. *Evolution and War.* Dutton, New York, 1915.
> An eminent biologist punctures the myth that war is an essential part of the process of human evolution.

MONTAGU, ASHLEY. *Man's Most Dangerous Myth: The Fallacy of Race.* 3d ed., Harper, New York, 1952.

——"The Nature of War and the Myth of Nature," *Scientific Monthly,* vol. 54, 1942, pp. 342-353.
> In the article and in chapter 12 of the book the author examines the common myths about the nature of "Nature" and the "inevitability" of war.

——*On Being Human.* Schuman, New York, 1950.
> A small book setting out some of the evidence which indicates that co-operation and not compe-

tition is the dominant regulative principle with which organisms are innately endowed.

MORGAN, A. E. "The Civilizing of Competition," *Main Currents in Modern Thought*, vol. 8, 1951, pp. 20-24.

The ex-chairman of T.V.A., in an article originally published in 1939, while retaining the term attempts to show how competition may be rendered more effective by making it more co-operative. The thesis is that co-operative competition is a social good.

MULLER, H. J. "Genetics in the Scheme of Things," *Proceedings* of the Eighth International Congress of Genetics, *Hereditas*, Supplementary Volume, 1949, pp. 96-127.

The presidential address by a great geneticist. Excellent, among other things, on the socio-biological aspects of man's evolution.

NASMYTH, GEORGE. *Social Progress and the Darwinian Theory*. Putnam, New York, 1916.

There is a brilliant introduction by Norman Angell. Nasmyth's excellent book is too little known. It is really a study of Social Darwinism, which ably shows that force as a factor in human relations is detrimental rather than beneficial. The book errs in claiming Darwin not to be a Social Darwinist. Highly recommended.

NEEDHAM, J. *History Is on Our Side*. Macmillan, New York, 1946.

A series of delightful essays, among the most important of which for our purposes is "The Gist of Evolution."

————*Time: The Refreshing River*. Macmillan, New York, 1943.

The theme of co-operation runs through this series of distinguished essays by an eminent scientist like a refreshing river. A book to read, study, and treasure.

NOVICOW, J. *La Critique du Darwinisme social*. Paris, 1910.

A very able criticism of Social Darwinism.

————*Mécanisme et Limites de l'Association humaine*. Paris, 1912.

An English translation by S. P. Otis and edited by C. A. Ellwood is available in the *American Journal of Sociology*, November, 1917. The emphasis is on co-operation.

PARRINGTON, V. L. *Main Currents in American Thought*. Harcourt, 3 vols. in one, New York, 1930.

See particularly vol. 3, part 2, chapter 1, pp. 189 *et seq.*, in which the impact of the industrial revolution, of biological thought, and especially of Herbert Spencer upon American thought is traced.

PATTEN, WILLIAM. *The Grand Strategy of Evolution*. Badger, Boston, 1920.

One of the finest expositions of the principle of co-operation in nature, by an eminent biologist who, after a lifetime of purely academic biology, "lets his hair down" in this volume. A very important book.

PERSONS, S. (editor). *Evolutionary Thought in America*. Yale University Press, New Haven, Conn., 1950.

Contributions by eleven authorities from various fields on the relation of the impact of Darwinism to American thought. Some of the contributors are separately listed in the present annotated bibliography. See Corwin, Faris, and Spengler.

POLANYI, KARL. *The Great Transformation*. Rinehart, New York, 1944.

One of the most important books of our time. It traces the effect of the developing market economy upon our developing conception of man. An absolutely fascinating book.

REDFIELD, R. (editor). "Levels of Integration in Biological and Social Systems," *Biological Symposia,* vol. 8, Jacques Cattell Press, Lancaster, Pa., 1942.

Twelve contributions by as many distinguished scientists from several different fields on integrative and co-operative processes in nature and in man.

REINHEIMER, H. *Evolution by Co-operation: A Study of Bio-Economics.* Kegan Paul, London, 1913.

An important study of co-operation as a factor in evolution. This work is far too little known, even though it was published about forty years ago.

——*Symbiosis: A Socio-Physiological Study of Evolution.* Headley, London, 1920.

By "symbiosis" Reinheimer means systematic biological reciprocity existing between organisms. His book, which is all too little known, presents a very stimulating and well-balanced account of the facts.

ROBERTSON, J. M. *A History of Freethought in the Nineteenth Century.* Watts, London, 2 vols., 1929.

A magnificent introduction to the background and the foreground of Darwinism through the brilliant discussion of the views of the principals and their forerunners. This is a fascinating book.

SHERRINGTON, C. *Man on His Nature.* Cambridge University Press, New York, 1st ed., 1941; 2d ed., 1951.
   One of the great books of this century by an outstanding physiologist. The concept of co-operation is beautifully considered, especially in chapter 12.

SIMPSON, G. G. *The Meaning of Evolution.* Yale University Press, New Haven, Conn., 1951; Mentor Books, New York, 1951.
   The best contemporary book on evolution, containing an admirable discussion of the co-operative factors in evolution.

SINNOTT, E. W. *Cell and Psyche.* University of North Carolina Press, Chapel Hill, N.C., 1950.
   Subtitled "The Biology of Purpose," this little book by a distinguished biologist argues that the purpose of the organism is to survive, and that co-operation rather than conflict is likely to further this purpose.

SMITH, KENNETH. *The Malthusian Controversy.* Routledge and Kegan Paul, London, 1951.
   A most valuable and interesting study of the predecessors of Malthus, and of his contemporary critics. Sound, scholarly, highly readable, and thoroughly critical of the Malthusian doctrine.

SOROKIN, P. A. *Altruistic Love.* Beacon, Boston, 1950.
   A study of American good neighbors and Christian Saints, in order to discover what the typical characteristics are of altruistic persons. Professor Sorokin is the pioneer student in this area.

———(editor). *Explorations in Altruistic Love and Behavior.* Beacon, Boston, 1950.

> A symposium by seventeen distinguished contributors representing a wide variety of fields. An important and highly readable volume.

———*The Reconstruction of Humanity.* Beacon, Boston, 1948.

> Particularly valuable for its practical recommendations for the altruization of mankind.

SPENGLER, J. J. "Evolutionism in American Economics, 1800-1946," in PERSONS, S. (editor), *Evolutionary Thought in America.* Princeton University Press, Princeton, N.J., 1950, pp. 202-266.

> One of the most valuable essays of its kind, brilliantly done.

STANDING, H. F. *Spirit in Evolution.* Allen & Unwin, London, 1930.

> An interesting book on the evolution and significance of love.

STAPLETON, LAURENCE. *Justice and World Society.* University of North Carolina Press, Chapel Hill, N.C., 1944.

> An exceptionally fine study of the concept of "Natural Law" and the significance of that concept for our own times. An indispensable and beautifully written book.

[TOWNSEND, J.] A WELL-WISHER TO MANKIND. *A Dissertation on the Poor Laws.* London, 1786. Reprinted in McCulloch—see above.

> This very rare tract deserves to be known to every well-educated person, for it exercised a very great influence upon nineteenth-century thought, through the alembic of such minds as Darwin's

and Malthus'. Townsend is the intellectual father of Malthus and the philosophical grandfather of Darwin.

TREMBLEY, F. J. "Evolution and Human Affairs." *Proceedings of the Pennsylvania Academy of Science,* vol. 23, 1949, pp. 181-195.
>   An invaluable discussion of evolution with particular reference to co-operation.

WEST, G. *Charles Darwin: The Fragmentary Man.* Yale University Press, New Haven, Conn., 1937.
>   By far and away the best biography of Darwin and his ideas. Before anyone is ever introduced to the theory of evolution I would make this book required reading. A fundamental book.

WHEELER, L. R. *Harmony of Nature: A Study of Cooperation for Existence.* Longmans, New York, 1947.
>   An interesting though uncritical work.

WHEELER, W. M. *Essays in Philosophical Biology.* Harvard University Press, Cambridge, Mass., 1939.
>   A series of very attractive essays by an eminent biologist. The theme of co-operation runs through most of them like the red thread of Ariadne.

ACKNOWLEDGMENTS

Thanks are due to the following publishers for permission to quote from the books published by them:

Richard G. Badger, William Patten, *The Grand Strategy of Evolution.* Cambridge University Press, Charles Sherrington, *Man on His Nature.* Columbia University Press, Th. Dobzhansky, *Genetics and the Origin of Species.* The Macmillan Company, S. J. Holmes, *Life and Morals.* Charles Scribner's Sons, Marston Bates, *The Nature of Natural History.* Charles Scribner's Sons, Marston Bates, *Where Winter Never Comes.* Harvard University Press, W. M. Wheeler, *Essays in Philosophical Biology.* Yale University Press, Geoffrey West, *Charles Darwin.* Yale University Press, G. G. Simpson, *The Meaning of Evolution.* W. W. Norton, R. E. Money-Kyrle, *Psychoanalysis and Politics.*

INDEX

bilizing Selection, n. 68, 76

SHERRINGTON, CHARLES, on co-operation, 64–65; 135

SIMPSON, GEORGE GAYLORD, on evolution, 51–55, 135

SOCIAL DARWINISM, and Herbert Spencer, 32–33, 94; 42

SOCIAL EVOLUTION, and biological evolution, 29–34, 73–75

SOCIAL LIFE, 44, 48, 73–75; W. M. Wheeler on, n. 67, 75; Th. Dobzhansky on, n. 67, 75; Darwin on, 74; S. J. Holmes on, 77–78

SOLITARY ANIMALS, non-existence of, 74–75

SOROKIN, PITIRIM, 12; writings, 135–136

SPENCER, HERBERT, and development of Social Darwinism, 32–33, 43; and "survival of the fittest," 57, 76, 108; and Gobineau, 98

STRUGGLE FOR LIFE OR EXISTENCE, 16–18, 74, 103–104; and, 25–26, 4.; *laissez faire,* n. 32; T. H. Huxley o. 38–39; Darwin defines, 45, 92–94, and mercantile civilization, 26–28, 45–49; G. G. Simpson on, 52–53; Th. Dobzhansky on, 56; K. Pearson on, 109–110; and co-operation, 70, 96; W. Roux on, n. 71, 77–78; and race, 89–91, 98; some representative statements on, 107–113; Theodore Roosevelt on, 110; Babbitt on, 112

SUMNER, WILLIAM GRAHAM, on competition, 108–109

SURVIVAL OF THE FITTEST, 11, 16, 17–18, 42, 87, 96–98; W. M. Wheeler on, n. 33, 43; group, 71–72, 104–105; S. J. Holmes on, 48; Geddes and Thomson on, 50; G. G. Simpson on, 51–55; phrase introduced by Herbert Spencer, 57; T. H. Huxley's criticism of, 57; C. D. Leake on,